Environmental Winds

Environmental Winds

MAKING THE GLOBAL IN SOUTHWEST CHINA

Michael J. Hathaway

UNIVERSITY OF CALIFORNIA PRESS
BERKELEY LOS ANGELES LONDON

University of California Press, one of the most distinguished university presses in the United States, enriches lives around the world by advancing scholarship in the humanities, social sciences, and natural sciences. Its activities are supported by the UC Press Foundation and by philanthropic contributions from individuals and institutions. For more information, visit www.ucpress.edu.

University of California Press
Berkeley and Los Angeles, California

University of California Press, Ltd.
London, England

© 2013 by The Regents of the University of California

Cataloging-in-Publication Data is on file with the Library of Congress.

ISBN 978-0-520-27619-2—ISBN 978-0-520-27620-8

Manufactured in the United States of America

22 21 20 19 18 17 16 15 14 13
10 9 8 7 6 5 4 3 2 1

In keeping with a commitment to support environmentally responsible and sustainable printing practices, UC Press has printed this book on Rolland Enviro100, a 100% post-consumer fiber paper that is FSC certified, deinked, processed chlorine-free, and manufactured with renewable biogas energy. It is acid-free and EcoLogo certified.

CONTENTS

List of Illustrations vii
Acknowledgments ix

Introduction 1

1 · Environmental Winds 8

2 · Fleeting Intersections and Transnational Work 41

3 · The Art of Engagement 73

4 · Making an Indigenous Space 116

5 · On the Backs of Elephants 152

Conclusion 185

Notes 189
Glossary 219
Bibliography 223
Index 251

ILLUSTRATIONS

Map 1.   Map of China *9*
Map 2.   Map of southern Yunnan *34*
Figure 1. Educated youth in Xishuangbanna *16*
Figure 2. The old traditional forester and the new social forester *20*
Figure 3. Auto repair class at Breakaway: A Women's Liberation School in Emeryville, California, 1973 with Chinese poster of female tractor driver *28*
Figure 4. Mao Zedong on the cover of *The Black Panther* newsletter, 1969 *30*
Figure 5. "Support America" poster from China *31*
Figure 6. Cultural Revolution woodcuts *32*
Figure 7. Prince Philip with Chinese scientists, 1986 *45*
Figure 8. Slicing wild banana stalks for pig food, 2001 *75*
Figure 9. Bamboo and grass houses in Xiao Long, 2001 *79*
Figure 10. Poster of protected animal species in Yunnan, 1999 *93*
Figure 11. Cover of WWF's *Agroforestry Handbook* *98*
Figure 12. Dai sacred landscape *139*
Figure 13. Hani sacred landscape *139*
Figure 14. City bosses pose with tamed elephants in Wild Elephant Valley, 2001 *180*

ACKNOWLEDGMENTS

Foremost, I would like to thank all of the many people I met in China, especially those who took care of my family in Xiao Long. I would also like to recognize my many friends and colleagues, some of whom have since passed away, that have worked, in various ways, to connect a love of the natural world with support for social justice in a challenging social context. This includes particular thanks to innovative pathbreakers such as Pei Shengji, Xu Jianchu, and Yu Xiaogang.

Among my network of close friends in Kunming and beyond, I would also like to extend thanks for warmth, support, and insight to Ulrich Apel, Graham Bullock, Cao Guangxia, Jim Harkness, Craig Kirkpatrick, Joseph Margraf, Nick Menzies, Min Guo, Bob Moseley, Ed Norton, Willem Quist, Tian Shitao, Vicky Yanzhen Tian, Jeannette van Rijsoort, Dan Viederman, Wu Deyou, Olivia Xue Hui, Xue Jiarong, Xue Jiru, Yang Bilun, Yang Yuancheng, Yang Yuming, David Young, Nick Young, Zhang Lianmin, Zhou Dequn, Yang Haiyu, Zhai Wen, and Zhang Hai. In the field Lu Wenhong, Yang Xueqing, and Zhang Hai provided dedicated and creative support.

During my doctoral studies at the University of Michigan, I relied greatly on my committee members. I am deeply grateful to Arun Agrawal, Gillian Feeley-Harnik, Conrad Kottak, Erik Mueggler, and Anna Tsing, for both pushing me and allowing me space to come up with something different.

My manuscript reviewers, among them Vanessa Fong and Anna Lora-Wainwright, pointed out gaps and encouraged me to strive for greater clarity of argument.

I benefited from excellent research assistance from a team of bright graduate students at Simon Fraser University: Karen-Marie Elah Perry, Grace Hua Zhang, Bardia Khaledi, and Dalia Vukmirovich. I would also like to extend my thanks for wonderful library support from Moninder Bubber, Rebecca Dowson, and Sonny Wong.

My studies of the links between China and the North American feminist movement were helped by Duke University archivist Kelly Wooten and student Michelle Lee. At the University of Michigan's Labadie Collection, Julie Herrada provided assistance. I relied on scholarship and advice from Carol Hanisch and Alice Nichols for help in understanding how these dynamic social worlds came together.

Much thanks to Judy Brandon for allowing her wonderful artwork to grace the cover of this book. I also appreciate Tom Dusenberry for permission to reproduce his photos of Yu Xiaogang, Julian Calder for allowing me to reproduce his photos of Prince Philip, David Young for allowing me to reproduce his image from a social forestry class in Kunming in 1995, Marisa Figueiredo of Redstockings for her assistance in finding images, Cathy Cade for helping me obtain permission for her photographs, and Charles Santiapillai for his photos of WWF's first elephant projects. I was lucky to work with a number of fantastic people at the University of California Press, who shepherded the manuscript through with efficiency and enthusiasm, especially Reed Malcolm and Stacy Eisenstark. In the final stages, the manuscript benefitted from the sharp eyes of Judith Hoover and Barbara Kamienski.

Feedback from audiences at the University of Wisconsin (Madison), University of Michigan (Ann Arbor), as well as several meetings of the American Anthropological Association, Association of Asian Studies, and the Society of Environmental History gave me new insights and approaches to this study. I also acknowledge generous support for this project from the National Science Foundation, U.S. Environmental Protection Agency, Social Science Research Council, Social Sciences and Humanities Research Council, Wenner-Gren Foundation, University of Michigan, and Simon Fraser University.

I am grateful for all of the helpful criticism and tough love and assistance from a wonderful group of scholars and friends: Bonnie Adrian, Nicole Berry, Alexia Bloch, Jeremy Brown, Kim Clum, Anita Crofts, Dara Culhane, Susan Erikson, Fa-ti Fan, Kim and Mike Fortun, Paul B. Garrett, Karen Hébert, Dorothy Hodgson, Milind Kalikar, Christina Kelly, Jake Kosek, Ralph Litzinger, Celia Lowe, Renisa Mawani, Emily McKee, Ed Murphy, Nancy Peluso, Ivette Perfecto, Stacy Pigg, Deanna Reder, Shiho Satsuka, Sigrid Schmalzer, Richard Schroeder, Elena Songster, Janet Sturgeon, Hannah Wittman, and Lingling Zhao. I would especially like to thank five people who were critical to making this book a reality: Juliet Erazo, Mel Johnson, Genese Sodikoff, Anna Tsing, and Kathy White. My parents, Walton and Peggy Hathaway, have helped in more ways than they know, instilling in me a deep love of the natural world and a strong sense of social justice. My biggest thanks, however, go to my family. The sharp wit of my son, Walker Hathaway-Williams, has helped me to navigate China, the United States, and Canada with much insight and laughter. My daughter, Logan Hathaway-Williams, has been a joy, and I look forward to bringing her to meet friends in China one day. Most of all, I thank Leslie Walker Williams, who has always been my sharpest critic and most generous supporter, the best partner in life I could ever imagine.

Earlier versions of several of these chapters have appeared in previous publications. Chapter 2 appeared as "Global Environmental Encounters in Southwest China: Fleeting Intersections and 'Transnational Work'" in *The Journal of Asian Studies* 69:2 (2010): 427–451. Chapter 4 appeared as "The Emergence of Indigeneity: Public Intellectuals and an Indigenous Space in Southwest China" in *Cultural Anthropology* 25:2 (2010): 301–333.

Introduction

IN THE SUMMER OF 1995, in an old greenhouse in the dry hills above Santa Cruz, California, a chance meeting would inextricably pull me into the world of environmentalism in China. My admiration of a rare orchid with delicate purple petals and intricate designs led to a conversation with a man named Karl Bareis, who told me that this specimen was brought from Southwest China to France in the late 1800s. I knew a bit already about Bareis, a renaissance man fluent in Japanese, a bamboo expert, and a cultivator of rare tropical fruit, and I was interested to learn that he himself had just returned from this region, along China's remote mountainous borders with Myanmar and Vietnam, on an ethnobotanical expedition. I had just finished my undergraduate thesis on global environmental politics in the Brazilian Amazon and was trying to find work in China. My wife, who had lived in Nanjing for a year during the mid-1980s, was eager to return to China. I too was deeply curious about "actually existing socialism" in China and what impact, if any, global environmentalism was having there.

Bareis talked excitedly about his trip, including how his group tried to track down "medicine men," as he called them, who knew how to find rare wild herbs and cultivate them in garden plots. He gave me the address of a fellow explorer, Xue Jiru (Hseuh Chi-Ju), a retired professor who first earned international fame in the 1940s.[1] Under great difficulties, Xue had collected botanical samples of a deciduous conifer tree, the Dawn Redwood (*Metasequoia glyptostroboides*), long thought by botanists to have gone extinct during the age of the dinosaurs (Hseuh 1985). Xue sent the specimen to Harvard's Arnold Arboretum, which welcomed his prized discovery. His pressed leaves and careful handwriting remain there to this day. I wrote to Xue and was delighted when, nearly seven weeks later, I received a warm reply, inviting me and my wife to spend a year teaching at his institution, the Southwest Forestry College in Kunming, the capital of Yunnan Province.

Our flight to Kunming had a one-day layover in Hong Kong, where I stopped in at the small, bustling office for World Wildlife Fund-China (WWF). I met the energetic American director and Hong Kong staff, who saw their job of promoting environmentalism in China as a daunting task. However, the director also suggested that "if there was something called 'environmentalism' anywhere in China, Yunnan was the place to find it." He contrasted Yunnan to Nanjing, where he had lived for years and where there seemed little love of the natural world. In Nanjing, one of the few remaining endangered Yangtze River dolphins just barely managed to survive in captivity, neglected in a pool filled with green algae blooms. In Yunnan, he said, there were still herds of wild elephants, which was a surprise to me. He was depressed about China's environmental future, but Yunnan Province gave him hope, both because it maintained a surprising diversity of flora and fauna and because of the conviction, energy, and abilities of a group of Chinese experts with whom he worked—people I would later get to know well.[2]

The next morning we landed in Kunming, one of the many cities in China barely known to the West but bustling with millions of people. Within a day my wife and I met Xue in his small apartment. He sported a blue "Mao suit" and an ebullient smile and was starting to stoop with age. He sat us down on the sofa and slid open the glass door under the TV set, grasping a thick glass bottle that was filled with red lycium berries and two dried geckos, their bodies lashed to a bamboo frame with red string. He poured us each a cup of strong spirits. We were introduced to Xue's son, a shy man in his forties, who had a single-minded passion for bamboo, and his granddaughter, then in high school, who was urged to "practice English."

Xue had learned English as a youth, and like many of his fellow scientists who matured during the 1930s and 1940s, China's "age of openness" (Dikötter 2008), his sensibilities were strongly influenced by that era. At the time he fostered strong international connections, hoping not necessarily to move abroad but to contribute to China's development. His son had grown up "between Russian and English," after China split with the Soviet Union in 1962 but before the beginning of the "reform period" in 1979, when China began to actively reach out to the capitalist world and English became important again.

Some hours later we left his apartment, slightly woozy from the reptilian brew and in possession of a beautiful book, *The Gaoligong Mountain Na-*

tional Nature Reserve, written in Chinese, produced by his research team as part of an ecological survey to create a new nature reserve (Li and Xue 1995). We soon became friends with Xue and a handful of his elderly peers, scholars who had come of age during a time of exciting cosmopolitanism and intense debate over China's future. They were friendly and active, optimistic for a new China that they said "cared about science, again." At first I didn't know what they meant by that "again." It took me months to realize that this and other sporadic remarks referred to an often tragic past, for many scientists had suffered greatly from the 1950s to the 1970s, accused of being "imperialist running dogs" or of practicing "bourgeois, capitalist science." Starting in the 1980s, however, many of these scientists were "rehabilitated," and their peers and superiors increasingly respected their past accomplishments and offered them new opportunities. Many older scientists began to encourage their children and students to pursue a scientific career, as it offered hope to advance themselves and the nation now that China was in a period of relative calm and stability. They understood that interest in the environment was now growing, and they were hoping to make of it what they could during their old age.

Some of these older scientists and their younger peers used a metaphor to describe these changes: "environmental winds" (*huanjing feng*). "Winds" (*feng*) was a word I often heard when people talked about the past, to describe times when political movements (like the Great Leap Forward or the Cultural Revolution) brought life-changing consequences.[3] The fact that they chose this particular term, typically used to describe powerful social transformations, to refer to changes associated with environmentalism indexes some sense of its intensity. We were in Kunming for barely a week before we started to feel the breadth and force of these environmental winds, although it took time to understand many of their effects. In part, the winds signaled a marked shift in past understandings and desires for a different future. Wastelands were no longer understood as places that had not yet been converted into agricultural fields but were seen, instead, as ecosystems and habitats. Similarly, swamps that had been drained were restored as wetlands. Scientists and others discovered undocumented wild animals and plants, classifying some as endangered and quantifying overall levels of biodiversity. People like Xue were asked to document and plan these kinds of changes, create maps for ecological protection, devise new strategies, and train nature reserve staff in new methods for collecting ecological data and protecting the reserves.

As I traveled out of urban Kunming and visited remote upland villages I soon realized that these environmental winds often brought dramatic and sometimes drastic changes for millions of rural Yunnanese. Unlike the scientists, very few rural people had a salary, and for almost everyone money was "hard to find" (*qian hennan zhao*). I was struck by the seeming similarity of urban life in 1995 between Kunming and major North American cities, but I found that lives in the countryside were strikingly different. Many of the rural citizens I met in 1995 (and lived with during fieldwork from 2000 to 2002) built their lives directly with their own hands in a way rarely done in North America. They built their homes out of trees they chopped down and sawed, and adobe bricks they dug and dried in the sun. They plowed fields with oxen, raised or hunted animals, cooked with wood they gathered from the forest, and grew almost all of their own grain and vegetables. I knew some "back to the land" people in California who tried to be self-sufficient, but in China such activities were not a personal idiosyncratic quest but a widespread social phenomenon. Such relative self-sufficiency was not an age-old practice but actually fostered by a "grain first" policy starting in the 1950s, when the Chinese state pushed rural residents to focus on grain and dismantled many rural cash-based craft specialties such as making paper or cloth from cotton and silk and turned these crafts into urban industries. For decades the government exhorted farmers to kill grain-eating pests such as waterfowl, clear land to expand fields, and use more chemical fertilizers and pesticides. The government largely ran on grain, not cash; rural residents paid taxes in bags of rice and wheat. In turn, the grain was key to international trade and rationed to urban residents, who in 1980 made up only 20 percent of China's population. Yet by the 1990s, after forty years of strong pro-agricultural policies that served to expand and intensify production in rural areas, farmers in some places saw new people coming to their villages, exhorting them to behave differently; officials showed them maps indicating that village forests were now requisitioned and placed under state protection, forest guards enforced these mandates, and police confiscated guns as part of new antihunting regulations. Village children learned different ways to think about nature and sustainability and criticized their parents' hunting and farming practices.

Back at the college in Kunming, my wife and I were quickly caught up in these winds, and our Chinese colleagues and friends asked us to assist them in many ways. We were there during a period later referred to by some Chinese experts as "the gold rush"—a time of burgeoning interest in Chinese cultural and biological diversity. These experts taught us a whole raft of acronyms, asked us to edit reports for environmental nongovernmental organizations, write proposals to European governments for hosting environmental projects, and coach them for visa interviews at foreign embassies so that they could attend scientific conferences. We were invited to participate in conferences and workshops hosted by the United Nations' Food and Agriculture Organization (FAO) and the Ford Foundation, groups promoting social forestry and investigating the gendered dynamics of subsistence activities.

At these conferences I saw how Chinese participants positioned themselves in relation to visitors from Cambodia, Vietnam, Nepal, Italy, India, England, and the United States. By organizing these conferences, they helped forge international connections and worked on translating issues and concepts, helping them to travel between these different places. The conferences, all using English as a common language, would often stimulate active discussion and debate over how to understand and reconcile what they often described as "foreign concepts" (*wailai sixiang*) and "Chinese conditions" (*zhongguo guoqing*).

This book explores the lives of these scientists and rural farmers as well as two other groups who were caught up in these environmental winds in Southwest China: state officials and expatriate (foreign) conservationists.[4] It shows how these actors and winds brought unexpected and transformational changes to the area's natural and social landscapes. These groups noticed and reached out to the environmental winds in different ways.

I pay particular attention to the world of the Chinese experts and how they helped make Yunnan into one of China's most important places in the global environmental ecumene. Others might regard these people as mere translators or even culture brokers, but I came to see them as much more. They did not just passively wait for or respond to international interest but sought out connections and helped to generate energy and interest. They helped connect Yunnan with international circuits by positioning their province as a space of great biological and cultural diversity. This was particularly difficult in the mid-1980s because outsiders often viewed China as an environmental lost cause—what Vaclav Smil's influential 1984 book described as a "Bad Earth," which was already plundered and had little wild nature left worth protecting.[5] In terms of culture, many saw China as a land of social homogeneity and uniformity; it was the land of the "blue ants," masses of peasants in blue Mao suits.[6] Despite these difficulties, Chinese experts sought out and built connections with people and organizations around the world to create new narratives and foster new relationships. Through multiple and varied efforts—conducting research, carrying out projects, writing and publishing in English and Chinese, mapping coordinates and producing data sets, creating lists of endangered plants and animals, recording indigenous knowledge, and hosting and initiating international conferences—Yunnan's experts were building their province into an important environmental hub.

Yet the experts did not accomplish this solely by themselves. They relied on villagers who maintained compelling indigenous knowledge. Their efforts were also boosted substantially through the persistence of valuable tropical rain forest and charismatic animals like the Asian elephant, which were used to show that China still possessed vibrant cultural and ecological diversity that was well worth protecting. They brought together these people and animals to foster a hub that linked them as part of an emergent global environmental network.

My conversations, observations, and interactions with Chinese scientists, officials, rural farmers, and expatriate conservationists led me to wonder how one might understand the often striking divergence and unevenness of such networks, both between and within countries. I knew that the texture of environmentalism was so different, for example, in the United States and Brazil (the two countries I was most familiar with), but I did not know how it could also be so different within one country. I found that a neighboring province, Guangxi, had similar levels of biological diversity, but it remained relatively ignored by this network. How was it that Yunnan went from a place that was stigmatized in China as backward, isolated and poor, and barely known abroad, to becoming a global hub of environmentalism? This led to two of the key questions that underlie this book: How are global connections made, and why do they happen so differently in different places? The Chinese metaphor of "winds" turned out to be particularly useful for grasping some of the answers to these questions.

Many scholarly and popular accounts portray globalization as flowing across the world like a flood, submerging local differences under a universal force (of Westernization or capitalism). I argue instead that there is no singular form of globalization that affects all places equally. What is often understood as "the global" is both quickly changing and highly diverse, with multiple globalizing logics, aims, and aspirations. This is precisely because every day there are many people in many places who are actively engaged in *making* what we understand as globalization. Globalization, then, is not the self-propelling movement of one form, logic, or modality but a place of articulation and human work that not only transforms what is often described as the global *but actually brings it into being*.[7] In my use of the Chinese concept of "winds," it should be stressed that such forces are not understood as natural and beyond human intentions but as created by people's efforts. This book explores how many efforts to forge and maintain connections are not actually successful, and if they are, they can become transformed into something quite different from their origins.

Since my initial year in China in 1995, I have been swept up in this world of Chinese environmentalism, and I have watched with great curiosity and interest how it has emerged and continues to change over time. Over six trips lasting a total of more than three years, I have conducted archival studies, carried out interviews, attended workshops, and spent eighteen months living in two rural villages that were part of international conservation projects. I continue to watch and participate in these emerging worlds, even now as I live in Vancouver, Canada. It turns out that Olivia Xue Hui, the granddaughter of Xue Jiru (the botanist who first welcomed us to Yunnan in 1995), came to Simon Fraser University (where I am now a professor) to earn a master's degree in environmental studies, conducted fieldwork on Tibetan pastoralists in Northwest Yunnan, and later worked for The Nature Conservancy in Yunnan. In China and Canada we have talked on more than one occasion about how much things have changed. New environmental issues are gaining center stage, and my peers in Yunnan now emphasize emerging controversies over dam construction, biofuel plantations, and forestry projects aimed at addressing carbon storage and global climate change. We know, however, that even while living in Canada, we are still part of Yunnan's changing environmental winds. Through the ongoing work of diverse organizations and individuals (including myself), the winds have changed and are continuing to change the lives of many people we know, as they are caught up in these transformations.

ONE

Environmental Winds

IN THE SPACE OF FORTY YEARS the People's Republic of China (PRC) went from being a harsh critic of Western environmentalism to what some see as an international vanguard, an "environmental state" (Lang 2002). In 1972 Chinese delegates at the world's first international conference on the environment, in Stockholm, refused to sign global legislation, arguing that pollution was a product of capitalism, not socialism (Tang 1972). By 2002, however, many outsiders praised the Chinese government's powerful and sweeping environmental laws in rural areas.[1] China enforced the world's largest logging ban, converted massive areas of agricultural and grazing lands to forest, and confiscated hundreds of thousands of guns as part of increasingly strict laws against hunting. The state is not the only actor; popular protests, now amounting to over 100,000 events a year, are increasingly expressed in environmental terms, and citizens rally to decry air and water pollution, as well as their relocation from massive dam projects (Economy 2004; Mertha 2008). Citizen complaints to the government about environmental issues rose tenfold between 1999 and 2009 (Moore 2009). A number of outsiders now describe China (using metaphors common a century ago) as "awakened" to the environment and regard this as an inevitable result of globalization.

Does China's recent attention to the environment demonstrate that, as the world is increasingly connected through globalization, all places are becoming more alike? Globalization is often thought to describe a "world becoming more uniform and standardized, through a technological, commercial, and cultural synchronization emanating from the West" (Pieterse 1995: 45). Many hold the related belief, expressed by the best-selling author Thomas Friedman (2006), that "the world is flat," as people everywhere have access

MAP 1. Map of China, showing Kunming (the capital of Yunnan) and Xishuangbanna (Banna).

to ideas, connections, and opportunities created by global systems such as the Internet. Most accounts of globalization take a bird's-eye perspective, which focuses on overall political trends or flows of global capital (Steger 2004). By looking closely at a social field like environmentalism and how it is playing out on the ground in one of China's most active regions—the southwest's Yunnan Province—this book offers a different interpretation.

This study of China's environmental politics provides a way to think differently about globalization, and in particular globalized formations. I use the term *globalized formations* where others might use the more common yet narrower term *social movements*. The most common image of a social movement is a street-based rally, where people fight to transform state policy, such as creating new civil rights laws. I use *globalized formations* to signal my interest in a broader constellation of social acts and spaces than what is often understood as a movement, which signals a more temporally and socially discrete set of events toward specific goals (Givan et al. 2010). My subject is to explore how new sensibilities are taken up, fought against, and transformed among a wider public. Examples of globalized formations include movements around gay, indigenous, and women's rights.² My analysis of globalized formations emphasizes the critical role played by ordinary people in what I refer to as "making the global."

My understanding of these processes has been shaped by my extensive and ongoing engagements with many people in Yunnan Province, where in 1995 I first worked, lived, and conducted research. I use oral histories, interviews, and archival research to take us back to the beginnings of international conservation efforts starting in 1986, when representatives of the World Wildlife Fund first came to inspect Yunnan's tropical rain forests and search for China's last herds of wild elephants. I explore the subsequent two and a half decades as Yunnan went from being a relatively unknown site for nature conservation to becoming a prominent and influential place for global environmentalism. By 2011 Yunnan was well known for its wide range of habitats, from lowland rain forests to rugged Himalayan peaks. It is highly mountainous and contains the headwaters of some of Asia's great rivers: the Yangtze (Chang Jiang), Mekong (Lancang), and Salween (Nu). It joined the list of the world's "biodiversity hotspots" and is now claimed as "arguably the most botanically rich temperate region in the world" (He and Li 2011: 484). Dozens of international nongovernmental organizations (NGOs), the World Bank, and the Asian Development Bank are deeply involved in trying to shape the management of these landscapes. Many would argue that Yunnan's inclusion within international conservation networks seems to provide evidence that globalization has flowed to even the most remote places.

Indeed when I arrived in Yunnan in 1995, I too understood environmentalism as a global flow that originated in the West and was now spreading throughout China, propagated by groups like WWF. When I started to teach at a forestry college, I found that my first-year students were often puzzled over what *environmentalism* meant. As I learned more about China's history, I began to understand why this might be the case. I grew up in the United States during what some called an environmental revolution, as exemplified by the world's first Earth Day in New York City in 1970. I was influenced by the legacies of Henry David Thoreau, John Muir, and Dave Foreman of Earth First!, a radical pro-wilderness environmental group. By high school I was a passionate environmentalist and worked on several campaigns to raise money to save tropical rain forests. Yet in 1970 China was in the midst of its own revolution, the Cultural Revolution, and some cities became combat zones where young Red Guards fought each other with grenades and tanks. Some of my students in Kunming were born in 1976, the year Mao Zedong died and the Cultural Revolution ended, and grew up after China's massive market reforms started in 1978. They had heard of environmentalism but were not quite sure what it really meant—unlike their teachers, who, I found out, had been engaging with it for years.

These teachers, Chinese scientists who were also my colleagues, referred to the rise of environmentalism as a wind (*feng*), specifically as "environmental winds" (*huanjing feng*), and it became clear that they were not simply accommodating global environmentalism as advocated by WWF.[3] Instead they were actively reshaping these winds conceptually and in relation to China's unique history.

The concept of the environmental wind was different from how I had previously understood the ways globalization works, particularly with respect to the metaphor of flows. When people speak of globalization as a flow, it suggests a force that emerges and spreads without human agency. In contrast, the Chinese view of winds as social formations, made and maintained by people, offered me a different perspective on how globalization happens. Through many discussions I had in Yunnan, it became increasingly clear that winds do not simply impact people; they are made, shaped, and transformed by people. The more I thought about the Chinese perspective on winds, the more I believed that it offered a different, more radical view of what we understand as globalization—one that moves us away from seeing it as a force that emerges by itself or is created solely by the efforts of a few powerful individuals and corporations, a "conspiracy of the rich." Rather, many people, from rural small-scale farmers to government officials, shape and make the global, but not necessarily with the same intent, capacity, or outcomes.

I also found the notion of winds an intriguing concept for thinking about power and the ways that groups can be forged and inspired by new political possibilities. Many analysts of China, from academics to journalists, regard politics and power as a top-down imposition, which ignores the ways diverse groups of people become caught up in new social formations. The metaphor of winds suggests that we cannot know what happens by only studying Beijing's political proclamations; instead it brings us into the lives of Chinese experts, rural activists, expatriate conservationists, local leaders, and all those who have a stake in what happens next. When winds are powerful, there are those who live in the full force of this power, those who live in the eddies, and everyone in between. But all are shaped by and all are themselves shaping the winds, regardless of their intentions.

Before I arrived in China, my readings in the social sciences prompted me to anticipate that rural peoples in the Global South, and particularly those described as indigenous peoples, were fundamentally resistant to external forces such as state mandates and global impositions. The notion of winds also challenged my expectations that rural people aim to live autonomous lives, that they strive to be free of interference from both state and global forces. The idea of winds refuses such schemes of fixed responses and clean divisions between local and global, and encourages us to look at how a range of people, including both urban experts and rural villagers, engage with forces in diverse and creative ways.

I also was interested in the dialogic and transformative aspect of winds. Like physical winds moving through a landscape, the movement of social winds is iterative; the social landscape is constantly shaped by and shaping the movement and power of the winds. When millions of people actively embraced China's Cultural Revolution—attending rallies, reading Mao's *Little Red Book*, traveling across the nation, and joining the Red Guards—this buoyed its strength. When others began to refuse some pervasive elements of the time, such as Red Guard groups who organized to oppose the rampant physical violence, these actions affected the force and qualities of the winds.

This metaphor of wind is not uncommon in China. Many elderly people described their lives as a dizzying series of winds: "Let a Hundred Flowers Bloom," the "Great Leap Forward," the "Cultural Revolution," and the "Opening of China." They described these changes as shifting winds rather than as concrete and predictable stages: things were quick to change, powerful, and then with little explanation gone but with ramifying and lingering effects. I was taught other words that describe situations similar to but different from a *feng*. The related term *re* (literally "hot" but figuratively a "fever") describes a situation closer to the English equivalent of "fad" or "fashion," which can be a large-scale social phenomenon but is often less all-encompassing and feels fleeting. The term *yundong* (mass campaign) describes government-led campaigns with discrete beginnings and endings that may not sweep people up. *Feng* instead refers to times of more diffuse but still notable changes that are deeply felt engagements. I build on this latter sensibility, as winds are not just terms for political events but structures of feeling that change what it means to live in the moment and create lingering effects.

After weathering these winds for years, many people said they had cultivated a heightened ability to detect shifting winds on the horizon. Their stories were full of accounts of trying to position themselves and their families to avoid the political purges and dangers that a wind like the Cultural Revolution could bring.[4] A wind could not only bring about threats; it could also provide potential advantages, depending on how one acted. Many people seemed alert to new winds, knowing that they were fleeting rather than permanent. A wind often started with little notice, and some rapidly became powerful. Just as quickly, a wind could change course or dissipate.

Let me provide another example of powerful winds that shaped Chinese history, winds that quite strongly shaped natural landscapes as well as social ones. The Great Leap Forward (1957–61) swept up millions of Chinese in enthusiastic all-out efforts to create backyard steel furnaces in order to quickly overtake England in steel production. Throughout the countryside, peasants scoured the land for iron ore and cut down millions of trees to fuel these furnaces, in some places leaving a wasteland of stumps. At the same time, they built a massive infrastructure of over forty thousand reservoirs and canals, significantly expanding the country's potential for agricultural irrigation.[5] Large-scale agricultural communes were quickly amalgamated, and leaders competed to produce previously unheard of levels of grain. In the midst of this rush to build socialism, things went seriously awry: grain yields were vastly exaggerated, large quantities of grain were siphoned off to feed city residents, and peasants neglected their fields, resulting in the world's largest human-caused famine. Approximately 30 million people died, and although estimates vary substantially, the vast majority of those who starved were rural farmers who lacked sufficient access to their own crops (D. L. Yang 1998; Thaxton 2008). The legacies associated with these winds are still felt as China continues to be powerfully shaped by the Great Leap's enduring ecological, cultural, and social effects, many of which are strongly debated today.

In China the idiom of *winds* has been used mainly to describe changes at local and national levels (like the Great Leap). I extend this concept to help us examine social change at broader scales—in this case, how the globalized formation of environmentalism is made and remade as it travels around the world. The ways that globalized formations work out in any place are strongly mediated by historical legacies and social landscapes. In Africa, for example, current forms of environmentalism engage with a legacy of nature conservation as a key European colonial intervention, and conservation remains more racialized and militarized in Africa than in any other region in the world. In China's case, specific Cold War tensions with the United States inflect how its government works with American organizations. This points to how "global" interactions can be more insightfully understood as particular transnational articulations. As well, globalized engagements are shaped both by how the Chinese state actually operates and how many foreigners view the Chinese state. Winds provide a way to think about how such formations come into being and how they are reciprocally remade through a study of particular encounters and interactions.

While environmentalism was not nearly as powerful as the Great Leap, starting in the 1980s this wind blew through Yunnan during a time of considerable social change.[6] In the beginning, many people, from high-level officials to college students to remote rural villagers, began to mull over the term *huanjing* (environment/environmental), which was gaining in prominence and power. It formed an umbrella concept, covering both older interests in soil erosion and water conservation and newer concerns about urban pollution and biodiversity conservation. By the 1990s millions of hectares were designated as nature reserves and fragile upland watersheds, and billions of yuan were spent to guard such reserves against local farmers, now regarded as threatening the land with their hoes, guns, and cows.

But as many understood, winds may change direction, and by the 1990s, a number of people, including some Chinese scientists and expatriate conservationists, began to challenge strict forms of nature conservation. They suggested that villagers should not be viewed as environmental adversaries but should be enlisted as partners, or at least stakeholders. Furthermore they argued that some of these communities were not made up of peasants but of indigenous people who possessed "indigenous knowledge" and "sacred forests" and were entitled to special rights under international law. Although this contradicted Beijing's insistence that all Chinese were equally indigenous and that no one in China deserved special rights, this community orientation nonetheless opened new spaces for differentiating between rural communities on the basis of indigeneity. It also unexpectedly provided scientists with a way of challenging mainstream development and conservation initiatives. These dynamics did not unfold simply as an extension of winds blowing from the West; they emerged out of unique histories and relationships among social groups, between people and nature, and between Chinese citizens and the state. This book explores how these winds caught people up and how people, places, and the winds themselves were changed unexpectedly in the process. It brings us into the lives of those individuals who not only encountered environmentalism but brought it into being in China. By exploring how these winds gained force and shaped Yunnan's social landscapes, this book addresses larger questions about how people in China, and elsewhere, are *making the global*.

WILDLANDS AND WILDLIFE IN THE PRC BEFORE THE ENVIRONMENTAL WINDS (1950s–1980s)

When environmental laws were first being made and enforced in Southwest China in the 1980s, they were particularly striking because such frameworks were so unfamiliar.[7] There were few legacies of restrictive nature conservation laws.[8] A brief tour through China before the environmental winds is instructive. In the mid-1950s there was a short-lived opening when biologists successfully lobbied to create nature reserves, but such efforts were largely abandoned within the decade. At that time, the entire country became swept up in campaigns, like the Great Leap, that encouraged the rapid expansion of agricultural lands, a process referred to as "opening wastelands" (*kai huangdi*). These campaigns intensified after famines in the early 1960s; officials exhorted rural Chinese commune members to clear forests, plow grasslands, and drain swamps to convert them into fields of grain. The slogan "Learn from Dazhai"—referring to a model commune that tirelessly converted sloping hills into irrigated, terraced fields—was promoted and widely emulated. In southern Yunnan thousands of youth from urban China came to slash and burn tropical rain forests and replace them with vast plantations of rubber and tea (see Figure 1).

Official attitudes toward animals during this time were basically utilitarian. Except for a few species that earned state protection, such as the giant panda, many animals were freely hunted, used for people's own sustenance, or turned into state goods. For example, China's top ornithologists, often trained abroad, were asked to determine which birds were farmers' "friends" and which were "enemies." China's most famous engagement with birds was called the "Destroy the Four Pests Campaign" (rats, flies, mosquitoes, and sparrows), started during the famine in 1960. In cities crowds of people gathered, chasing flocks of sparrows from branch to branch, yelling and banging pots and pans. The exhausted birds fell from the sky and were collected by organized groups, who also tallied the bodies of the other "pests" brought to

FIGURE 1. Educated youth in Xishuangbanna. Image and text from a Chinese government English-language promotional book, *Learning from Tachai*. In the original publication, the caption below the image reads: "Chu Ke-chia (*first right*) working together with commune members. Upon leaving middle school in Shanghai Chu Ke-chia settled in Menglun People's Commune, Mengla County, Yunnan Province. Many city youths, like him, are answering Chairman Mao's call to live and work with the former poor and lower-middle peasants in building socialist new villages. Chu Ke-chia, who was made deputy secretary of the commune's Taka Brigade Party branch, has been elected concurrently an alternate member of the Party Central Committee."

them. Later, scientists argued that sparrows ate crop-damaging insects, and the sparrows were relabeled "friends" and removed from the list of pests. These accounts, which moralize the past through reference to ecology, are now commonplace in China and abroad as a way to invoke the tragedies of the Mao era.[9]

In the 1950s biologists published a series of books intended for commune leaders to teach farmers about wild animals and their "rational exploitation" (*heli kaifa,* a term with a positive connotation). Some birds, such as the black-necked crane, historically considered sacred in China (Matthiessen 2001), were killed for their down feathers, which were sorted into three grades and mainly shipped abroad.[10] Biologists were sent to explore western China, most of which remained a great unknown to those based in Beijing. Other waves of experts, including geologists, linguists, and anthropologists, were sent to map, classify, and develop rural areas economically and politically (Kinzley 2012; Arkush 1981).

By the early 1980s, during China's "Reform and Opening Up" (*gaige kaifang*) period, officials dismantled communes and divided up agricultural and forest lands among families, for the first time in over a generation. It was hard, though, for people to put much faith in the permanence of such land divisions after a history of tumultuous change; for example, the ownership of trees had officially changed hands up to seven times in the previous decades (Liu 2001). This transition period stimulated what is sometimes called the third of the "Three Great Cuttings" (*san da fa*) of trees, after the Great Leap Forward and during the Cultural Revolution (Hyde et al. 2003). Villagers built new houses, stockpiled lumber in ponds, and sold standing forests to entrepreneurs who sprung up in the new cracks that opened in the state-controlled economy.[11]

When the government began enforcing environmental laws in the late 1980s, villagers often regarded these changes with a mixture of shock and resignation. The laws greatly restricted villagers' capacity to farm, hunt, collect firewood, and take care of their animals (sheep, cows, goats, pigs, mules, yak, and horses) by taking them to graze, collecting wild foods, and gathering bedding material for their pens, such as pine needles and oak leaves. Together these tasks formed the core of the vast majority of rural people's livelihoods, providing goods for self-provisioning and the market. Many had few alternatives for supplementing household income, as opportunities for paid work were rare. Moreover villagers were shocked to find that the new law positioned them as criminals. For decades they had been regarded as China's vanguard, the key force of the socialist revolution, and the providers of critical grain supplies. They had followed patriotic mandates to build rural China into a land of productive agriculture, and now these landscapes were being dismantled or abandoned for reasons that were often unclear. Even worse, such rules distanced them even further from their status as China's revolutionary masses (*geming qunzhong*), a position that they had already largely lost by the early 1980s, when China turned away from socialist orientations and toward a market economy. The new regulations were part of a growing official and urban view that rural people were the presumed enemies of wild animals, forests, grasslands, and wetlands, all of which were now protected by state employees (Williams 2002).

Environmentalism in China acted more as winds blowing through an intensely variegated landscape than as a singular, homogeneous flow that submerged local differences. As drastic as these changes seemed to me, many said that this was the condition of life; these environmental regulations were just another in a series of powerful and yet unpredictable winds. Once, when I was asking a woman about her labors during the Great Leap Forward, her thirty-something daughter piped in, both serious and joking, "You see how quickly things change? We might as well make use of our family fields. Who knows, next year, maybe we'll go back to communes." Such regulations were enforced in a highly uneven manner, so that people in cities and regions devoted to high-yielding paddy rice agriculture often experienced few environmental constraints. However, for those living in places regarded as ecologically valuable, and especially places designated as nature reserves, these laws could be quite powerful.[12]

Yet by the late 1990s China started to enact forceful policies that covered vast areas, precipitated after the Yangtze River flooded its banks in 1998, devastating a huge region of 25 million hectares and killing thousands (Lang 2002).[13] Chinese officials blamed upstream logging, farming, and grazing for causing the catastrophe and, in response, started to ban agriculture on slopes with greater than a 25-degree angle throughout the massive uplands watershed of the Yangtze River, an area more than twice the size of California.

This policy was especially harsh for farmers in Yunnan, who often plant crops on steep slopes in a province that is 95 percent mountainous. The ban also caused havoc for local governments, some of which earned 80 percent of their revenues from taxes on logging. These local governments, cut off from logging revenues, struggled to find new sources of income, such as promoting ecotourism or hunting wild matsutake mushrooms for export to Japan. Villagers were told to stop farming sloping lands, restore grasslands, and plant trees. Engineers were hired to break dikes and restore wetlands, destroying families' fields and crops. A hunting ban was followed by a large-scale campaign to confiscate guns throughout western China. Villagers I lived with were deeply impacted by losing their guns, which had been a critical part of everyday life for over a century. They had relied on guns to hunt, protect themselves against bandits, and scare away crop-raiding animals (including elephants). Indeed some upland families credited their guns with keeping them alive during the Great Leap Forward famine by making it possible for them to live on wild game. It is thus not surprising that handing over their guns to the police was felt even more intensely than other laws that restricted their use of forests for planting crops or gathering fuel wood.

SHIFTING WINDS: THE COMMUNITY-BASED APPROACH AND THE RISE OF INDIGENEITY

By the late 1990s a number of experts, conservationists, and development staff had begun to challenge such a hard-line position and advocate for working with communities in ways that appreciated their knowledge and, to some degree, promoted their rights. This new approach was not always embraced,[14] but it provided new techniques, trainings, and funding packages, as well as a critical vocabulary and webs of support. It created new opportunities for social scientists to work together with natural scientists, foresters, and development agents; to visit villages and build new networks, such as community forestry associations; to forge links between rural villages and urban centers, and between people in China and throughout the world. Yunnan, in fact, became an early leader in China's experiments with community-based techniques, and it produced several activists who later challenged international conservation NGOs working in Yunnan that were slow or inconsistent in applying a social justice framework. Kunming was becoming one of China's centers for international conferences on environmental topics and a major site of translation between foreign and Chinese agendas, and many livelihoods began to depend on these relations.

The overall change was dramatic. Those who followed new community-based models of development saw local people as having knowledge and experience that made them valuable partners for development experts.[15] The larger aim was often to recognize communities' history of land use and build collaborative relationships. This involved decriminalizing, legitimizing, and even encouraging some practices that had been frowned upon during the period of strict conservation, such as the collection of herbs and mushrooms from state-claimed forest lands.

As well, many others promoted "social forestry," which transferred forests' management from state control to villages. Whereas earlier models

	The 'Old' Traditional Forester 传统林业工作者	The 'New' Social Forester 新社会林业工作者
1.	Rural people have little or no knowledge of tree planting. 村民很少或没有植树知识。	Rural people have much knowledge about the use and care of trees and shrubs. 农民有许多利用和护林的知识。
2.	Rural people exploit nature beyond the limits of sustainability. 农民利用自然资源超出持续性极限。	Rural people try to care for nature but are often forced by outside factors to over-exploit it. 农民企图保护自然,但由于外界因素所逼而对自然进行过度开采。
3.	We should educate rural people about the importance of nature to their daily lives. 我们应教育农民使其认识到自然资源对其生存的重要性。	Rural people are very aware of the importance of forests in the local environment. 农民十分清楚森林对当地环境的重要性。
4.	Since they lack knowledge, they must be taught. 由于他们无知,需进行教育。	They need encouragement and technical help to do what they think is best. 他们需要鼓励和技术帮助去做他们认为最合适的事。
5.	If we motivate them properly, they will be willing to plant trees. 如鼓励得当,他们将乐意值树。	Planting is just one option. There may be good reasons for not planting. 植树仅是一种选择,他们可能有合理的不值树原因。
6.	If the right incentives are given, they will participate effectively. 如给他们适当刺激,他们将踊跃参加。	Rural people should be assisted to identify and design the activities which are best for them. 应帮助农民确认和设计使他们最为得益的活动。
7.	They tend to see woody vegetation as competing with other agricultural uses, so it should be separated as much as possible. 他们倾向于把木本植被看作与其它农业用地竞争,因此两者应尽可能分开。	Rural people, in principal, consider trees to be an integrated part of land use. But modern agriculture ignores, so separates, trees. 从本质上讲,农民认为树林是土地利用的一部分,但现代农民忽视因二分离树木。
8.	If people plant trees, the benefits will be enjoyed by the whole community. 如果人们植树,其利益属于整个社区。	The issue of who benefits depends on ownership, user rights, and control. 谁将受益取决于所有权、使用权和控制权。

Adapted from 'The New Forester' by Berry van Gelder and Phil O'Keefe, 1995

FIGURE 2. The old traditional forester and the new social forester. From a social forestry workshop in Kunming, 1995. Courtesy of David Young.

assumed that villagers' ecologically damaging practices arose from ignorance, development workers influenced by community-based sensibilities searched instead for what was referred to as indigenous or local knowledge. Such transformations in how nature conservation was carried out were not just happening in China but were part of a vast international shift. It would be easy to interpret this as just another instance of the flow of Western trends. However, I show that such changes did not result from a paradigm shift invented in Western headquarters but arose and were motivated by the cumulative experiences of millions of people who were affected by conservation efforts, both rural villagers and urban experts, including those in Yunnan. These cumulative experiences were generated by the struggles between field staff and villagers in implementing a hard-line, strict conservation approach. The many tensions that arose attracted the attention of an increasing number of critics, both inside and outside conservation organizations.[16] Critics described this strict approach as "coercive conservation" (Peluso 1993: 199) and argued that such practices were not only socially and morally unjust but actually undermined effective conservation in the long run (Peet and Watts 1996; West and Brechin 1991; Brockington and Igoe 2006).[17]

Initially I thought that this new community-based perspective represented a paradigm shift, both because the growing strength of that perspective seemed to ensure its dominance and because its conceptions of nature, local people, and governance were so different from earlier approaches. Yet it became increasingly clear, in both academic and policy debates, as well as through my own observation of environmental politics in China, that a community perspective did not completely displace earlier winds that were informed by coercive conservation[18]—just as the environmental winds themselves had not totally replaced earlier mandates to expand and modernize agriculture. Thus I began to see the changes that have occurred over the past three decades less as a paradigm shift and more as a series of shifting winds—forces that changed in tempo, direction, and intent.

Overall I argue that these shifting environmental winds neither were totally invented in China, nor did they arrive fully formed from the West. People in China, from scientists to villagers, officials to expatriates, were not merely affected by these winds; rather, as they observed, reached out to, and worked with these winds, people both made and transformed them. A diverse set of individuals came together in a number of ways: conversing by phone, fax, letter, and report; meeting in elegant hotel lobbies, austere government offices, and village rooms with hard-packed dirt floors. Together they were creating new ways to understand and govern landscapes. In the next section I introduce some of the dominant theories for understanding globalization and show how what I found in China suggests something quite different.

GLOBALIZATION AND GLOBALIZED FORMATIONS

It is said that we live in a globalized world, but what does this mean? In the 1950s scholars began using the term *globalization*, mainly to describe the internationalization of business (Simpson and Weiner 1991). Since that time the concept has expanded, and now it often refers to a strong break from an earlier world purportedly composed of relatively isolated places and culturally distinct peoples, where individuals, goods, and ideas circulated much less and at a slower pace. Globalization has become an increasingly important topic in academic and popular culture, often used as both description and theory to explain everything from the spread of a singular economic model (neoliberal capitalism) and a singular subjectivity (the neoliberal subject) to the flow of globalized cultural formations of all sorts, from feminism to human rights and environmentalism.

Although most accounts of globalization explicitly focus on economic and political realms (Steger 2004), I am more interested in cultural globalization, an interest shared by many scholars, including Arjun Appadurai, Ulf Hannerz, Margaret Keck, and Kathryn Sikkink. Appadurai (1996) challenges notions that the world is increasingly "Westernized," and instead describes five emergent global "scapes" in which many people across the globe participate, including mediascapes (the distribution of the capabilities to produce and disseminate media) and ideoscapes (the ideologies of states and counterideologies of movements, around which nation-states have organized their political cultures, that circulate around the world). Hannerz (2002) explores the movement of cultural products, emphasizing that a few countries, such as the United States, are especially powerful exporters. Although I don't totally disagree, I suggest we might use different theories to look at these dynamics; *importing* and *exporting* may be more useful terms for describing the trade in commercial goods and less useful for mapping transnational cultural movements, in which ideas and strategies are substantially modified, transformed, and recirculated. Such cultural movements are examined in Keck and Sikkink's important book, *Activists beyond Borders* (1998), which contributes to our grasp of globalized formations such as feminism and environmentalism, showing how activists create networks and make ideas and strategies travel. I am inspired by their work but also interested in how new social forms are emerging, without presuming that transnational activism represents the spread of political and social norms.

Additional scholars in many disciplines, from English to anthropology, geography, and critical development studies, have critiqued some of the main tenets of globalization theory, such as that the world is increasingly undergoing not only economic but cultural homogenization based on a Western model (Foster 2008; Ong and Collier 2005; Inda and Rosaldo 2002; Haugerud 2005). Many are wary of the way this belief resonates with enduring notions of Western superiority and of an inevitable "march of history" from barbarism to civilization—ideas used to justify European imperialism, colonialism, and neocolonialism (Ashcroft 2001; Robertson 1995). A number of historians question mainstream globalization theories, showing that movements of people, economic goods, and ideas are neither new nor of unprecedented importance (McKeown 2007).[19] Others suggest that globalization is not as much flat and evenly distributed as it is "lumpy" (Cooper 2005) or exists as a "network of points" (Ferguson 2006), or is built through "trajectories" (Hart 2002) or labor-intensive social imaginaries (Ho 2003). Still others point to places, where global capital does not flow and where states do not attempt to govern their subjects, that they call "black holes" (Trouillot 2003), "ungovernable spaces" (Watts 2003), and "zones of abandonment" (Biehl 2005).

Despite such insights, scholarship on globalization tends to gravitate toward or emphasize one of three different tropes that characterize how people encounter globalization: *resistance, accommodation,* and *localization.* Resistance models emphasize how subalterns (groups in positions seen as less powerful, such as the poor, women, and indigenous peoples) resist forces of domination. Many accounts, especially in the Global South, rely on the now well-established concept of resistance to explain subaltern actions and orientations. For example, a number of scholars argue that indigenous groups "resist neoliberalism" (Stahler-Sholk et al. 2007), and a prominent scholar on international indigenous law argues that "indigenous movements are a form of resistance to modernization and globalization" (Kingsbury 1998: 421). I would argue instead that indigenous social movements are very much a part of our contemporary global landscape. These notions of opposition draw on a longer history of accounts of peasant resistance against states, colonial forces, and local elites; one of the more widespread stories describes women in India hugging trees to prevent them from being cut down by outsiders (Guha 1990; Scott 1985; Shiva 1988).[20]

In contrast to an emphasis on resistance, a number of scholars from several frameworks emphasize how groups are brought within overarching structures, whether or not they resist them. I refer to such studies as "accommodation."[21] Some advocates of such studies argue, for example, that over the past century most people on earth have accommodated capitalist mandates, including participation in wage labor and a money economy. In this case, resistance may be a temporary phenomenon but is ultimately supplanted (Taussig 1987). A quite different body of scholarship builds on Foucault's notion of governmentality, which explores the "art of governing," the ways populations are conceived of and managed, and the ways subjects fashion themselves (Rose et al. 2006; Luke 1999).[22] From this perspective, power is not an external force that already made subjects resist; power circulates through subjects and is part of subject-making itself. Even though they are strikingly divergent, each group of scholars tends to emphasize and even presume the success of subject-making, whereby people accommodate and become, for example, capitalist or neoliberal subjects.

Localization theories, in contrast, emphasize cultural difference and stress the necessity of creative work to bring global goods and concepts into particular worlds. Some localization scholars challenge the very categories of global and local—a presumption that the world can be neatly cleaved in two—and many question narratives of active global forces and passive local recipients. Instead they show how even apparently similar global phenomena, such as eating McDonald's food or watching the TV soap opera *Dallas*, are made locally meaningful in diverse ways (Watson 1997; Liebes and Katz 1990).[23]

As was mentioned earlier, scholarship on globalization tends to emphasize one of three different tropes that characterize how people encounter globalization. Yet in trying to understand how globalized environmentalism played out in China, I realized that, as different as these models are, they are limiting our understanding, as they tend to encode four central assumptions that permeate work on globalization. They tend to assume (1) that the global (in various forms, such as science, capitalism, and neoliberalism) is something forged in the West; (2) that global forms are relatively fixed or change in stages, such as capitalism moving from Fordist to flexible forms;[24] (3) that global formations flow only in a one-way trajectory, impacting only local cultures; and (4) that global encounters take place either between a global and a local group or between two groups regarded as having known identities and interests. The way that I saw environmentalism unfold in everyday lives, however, belied these models and presented a picture that was far more reciprocal and transformative than the models lead us to conceive. This led me to consider more seriously the Chinese metaphor of winds as a concept for thinking about globalization. As much as studies of localization can explore with great precision how novel forms are incorporated into daily lives (and thereby avoid the trap of arguing that people either accommodate *or* resist), they assume a unilinear direction of flows. They are thus unable to inquire into the very question of how "the global" and associated social formations are made and remade.

My perspective on globalization is informed by scholarship that seeks to "provincialize Europe" by elaborating alternatives to the dominant accounts of modernity, nation formation, and social development that are often Eurocentric (Chakrabarty 2008). Even in many accounts of globalization that explore diverse settings, scholars' arguments that, for example, Indonesians "dub" Western media (Boellstorff 2005) or Africans "cannibalize" Western ideas both reinforce the idea of a one-way flow and may reify the notion that projects are "Western" or "local." I join critics of studies on the economy, colonialism, and global development who point to the problematic prevalence of asocial, all-powerful, machine-like forces that are said to explain social change (Gibson-Graham 1996, 2006; Watts 2001). For example, colonial studies moved away from a primary focus on the colonized and assumptions "that what it meant to be European, Western, and capitalist was one and the same, to a more nuanced approach that questions the dichotomy of colonizer/colonized and examines instead interactions of engagement, intimacy, inequality, and opposition" (Cooper and Stoler 1997: viii). Such studies, then, explore colonial engagements not just as Western impositions but as activities that made and remade capitalism and the "West" itself (Mintz 1985; Blaut 1992).

My analysis incorporates these insights, building on studies of global encounters using what one could call a "world-making" approach. This includes, among others, work by Mei Zhan (2009), Lieba Faier (2009), Anna Tsing (2005), Sasha Welland (2006), Rebecca Karl (2002), Renya Ramirez (2007), and Celia Lowe (2006).[25] One of the fundamental features of this approach is that it expands the feminist insight that identities are formed relationally through interaction, and it applies this insight to the question of larger social formations, such as environmentalism and globalization. World-making challenges the premise that groups have singular and fixed identities and mechanically pursue interests that are already known and congealed. Instead it views identities and interests as constituted through social interaction.[26] Zhan's (2009) work, for example, argues that "Traditional Chinese Medicine" (TCM) was not fully formulated when it arrived in California in the 1980s. Other scholars might assume that TCM was "Chinese" and became "Americanized" in the United States. Instead Zhan shows how the cultural engagements of a wide range of people (including patients, practitioners, critics, and advocates of various ethnicities and nationalities) forged relationships, created exchanges, and made negotiations that crossed the Pacific, shaping TCM on *both* shores. This is thus more than a typical localization study, as it takes a dialogic approach to tracing ramifications rather than proposing an impact model.

EAST WINDS AND WEST WINDS IN A GLOBAL WORLD

Most people in China use the concept of winds to refer to powerful, domestic, state-led events, but I began to understand how the metaphor of winds can work beyond national borders, as a way of thinking about how social formations become globalized. One of the few and now famous references to global winds was during the Great Leap Forward when Mao triumphantly declared in 1958 that the "East Wind will defeat the West Wind." Mao borrowed the term East Wind (*Dong Feng*) from ancient Chinese history to stand for socialism (led by China and the Soviet Union), while the West Wind (*Xi Feng*) stood for capitalism (led by the United States and England).[27]

Although I use the concept of winds differently, one can adopt it to rethink dominant assumptions about how globalization works. For example, most accounts of China regard the Mao era as a time of autarky and inward self-sufficiency, when China severed its existing international connections, and the "bamboo curtain" went up (e.g., Wong and Han 1998).[28] Most descriptions of China's spectacular rise to a place of global influence begin their narrative after Mao's death and China's "opening" in 1978 (e.g., Greenhalgh 2010). According to this perspective, China was on the receiving end of globalization until the 1980s or 1990s, when it began to have global power and reach. Although it may be true that in global economic terms, Mao-era China (1949–78) was relatively insignificant, I suggest that China played a significant role in inflecting globalized formations, a role that is largely forgotten today.

I now offer two brief examples, one "global" and one "local," to demonstrate more concretely how I understand the way winds work, travel, and transform. The first example looks at "America's Cultural Revolution" in the 1960s (Kimball 2001) to consider the importance of China and Third World liberation movements in inflecting the goals and language of the women's and civil rights movements, which are often regarded as largely domestic affairs.[29] The second example is more familiar, wherein a Western project is being carried out in a developing country; this is the more typical conception of how global flows work, from "the West to the rest."[30] In this case, I will take you to Yunnan Province, where WWF's tropical rain forest conservation project was being evaluated by a team from the European Union.

The Winds from the East: China's Role in the Global 1960s

Although few Mao advocates actually traveled from China to North America or Europe, images and stories of China's liberation began to change the way Westerners thought about power, gender, and society in the 1960s.[31] In North America it was a time of challenging authority, uprooting inequities, and experimenting with radical egalitarianism, and it was a time deeply influenced by news of China's Cultural Revolution. Several social movements, such as women's rights and civil rights, owe much to a number of Third World struggles. In particular, China provided critical tools and excitement as a model of liberation and revolution for feminists and black nationalists (Ho and Mullen 2008; Prashad 2002), though this is rarely acknowledged.[32]

Right from its inception, the People's Republic of China promoted itself as a global leader of women's liberation; women's rights advocates around the world invoked Chinese policies to argue for enhanced rights in their own countries. In 1950, the first year of Mao's rule, China criminalized forced marriages and legalized divorce, far earlier than most countries. In North America feminist practices such as consciousness-raising sessions drew directly on Chinese models of politicization, borrowed from widely popular books like *Fanshen* (Hinton 1966). According to Hinton, *fanshen*

FIGURE 3. Auto repair class at Breakaway: A Women's Liberation School in Emeryville, California, 1973 with Chinese poster of female tractor driver on the left. The teacher is still an auto mechanic in San Francisco. Photograph by Cathy Cade/Bancroft Library, UC Berkeley.

means "to turn the body," or "to turn over." To China's hundreds of millions of landless and land-poor peasants it meant to stand up, to throw off the landlord yoke, to gain land, stock, implements, and houses. But it meant much more than this. It meant to throw off superstition and study science, to abolish "word blindness" and learn to read, to cease considering women as chattels and establish equality between the sexes, to do away with appointed village magistrates and replace them with elected councils. It meant to enter a new world (vii).

From this book and others, feminists in North America learned that in China many people were encouraged to "speak bitterness" (*su ku*), to speak of their suffering and hardship in public forums. North Americans also borrowed this technique of collective discussion to share stories of their oppression as individuals and as women, to understand the social nature of their experience. Like the many study groups in China, North American feminists organized their own groups to study and discuss readings.[33] Slogans like "The personal is political," which changed the ways people understood, experienced, and engaged with what would count as "political," were inspired by readings from China (Hanisch 1970, 2006).

There was also much crossover between work in feminism and work in civil rights, and both groups drew on China's inspiration. Mao's *Little Red Book* sold well not only among socialist feminists but also among Black Power advocates, and it was a common sight in Harlem and the San Francisco Bay Area in the 1960s (Kelley and Esch 1999). Members of the Black Panther Party often quoted from Mao; by the late 1960s, when they began to shift from a nationalist to an internationalist movement, they required all members of the party to read Mao's book, thereby forging what some call "Black Maoism" (Austin 2006; Kelley and Esch 1999). Powerful leaders like Malcolm X invoked China as a positive example, stating that the Chinese had killed off their landlords, whom he referred to as their "Uncle Toms."[34] Mao, in turn, gave several talks about American racism, and China became a destination for radical blacks, including W. E. B. DuBois and the Black Panthers' Huey Newton.[35] The Black Panthers were part of a larger group sometimes referred to as North America's "Third World Leftists" (Young 2006), many of whom began to see the Vietnam War as a parallel to the creation of Third World conditions for ethnic minorities in the United States.[36] While China played an influential role in feminist and civil rights social movements in North America, it did not do so by using much material aid like funding or guns; instead China reached out with images, text, and speech.[37]

China's influence in North America was not singular but combined with many others, including the work of Frantz Fanon, the radical critic of French colonialism. North Americans worked hard to translate and transform these winds, including Maoism, into viable strategies and ways of talking and doing that brought something new while being reconfigured to make sense in their time and place.

North American radicals were also deeply inspired by and connected to the strong anticolonial movement in Africa, where, unknown to many, China was playing a strong role, not just conceptually but also materially. Starting in the 1950s, thousands of Africans traveled to China for advanced training in medicine, engineering, and even guerrilla military tactics (Brady 2003). In 1962 the East Wind bifurcated when the alliance between China and the Soviet Union turned antagonistic, unleashing a three-way "scramble for Africa" among China, the USSR, and the United States. Through the 1960s and 1970s, during a supposedly autarkic era, Chinese activists traveled to Africa, where they staged plays arguing that Africans and Chinese, as "people of color," should stand together against the United States and Soviet

FIGURE 4. Mao Zedong on the cover of *The Black Panther* newsletter, 1969. Courtesy of David Hilliard.

Union (Larkin 1971). China was a strong supporter of independence movements in Africa, in particular anticolonial struggles against England, France, and Portugal, sending shiploads of guns and other weapons,[38] as well as being heavily involved in Africans' political and economic development. In other words, China not only brought colonialism to an end more quickly in Africa, but it also fostered the spread of Maoist socialism.[39]

China's anti-imperialist efforts fostered winds blowing throughout the world, shaping political landscapes on several continents. China's influence, direct and indirect, helped stimulate the particular social conditions of the global 1960s, which included a desire for shared internationalism; a critical skepticism toward big government, big business, and authority; and a sense

FIGURE 5. "Support America" poster from China. The original caption reads: "Resolutely support the American people in their resistance against American imperialist aggression in Vietnam." Landsberger Collection. Courtesy of International Institute of Social History, Amsterdam.

of utopian hope (Wolin 2010).[40] These sensibilities connected people who would help establish environmentalism as a new social formation, symbolized by the first Earth Day in 1970.[41] Although environmentalism, like feminism and civil rights, is often assumed to have been invented in the West and diffused to the rest of the world as part of a global flow, these formations are in fact indebted to winds deeply influenced by China.[42]

Of these three movements, feminism is the one that drew most directly on Chinese concepts, strategies, and inspiration, yet very few studies, with

FIGURE 6. Cultural Revolution woodcuts of a female scientist and a tractor driver on stationery used by Redstockings, a socialist feminist organization based in New York. Courtesy of Redstockings.

only a few exceptions (Ross 2010; Lieberman 1991), mention, let alone delve into, China's legacy of inspiring, indeed making what we think of as "Western" feminism. This Eurocentrism is typical of many studies of globalization, which assume that the West is the sole innovator of social change.

Ironically it was not while I was in China that I first learned about the historical role of the East Wind on the West's liberation movements, and indeed many of my Chinese peers also used the term *Western feminism,* as though it was created solely in the West. Instead I found these transnational links in a dusty archive in Ann Arbor, Michigan. There, looking through a collection of materials created by North American radicals, such as feminist, anti-imperialist, and Black Power activists, I was shocked to see so many texts and woodcuts (the popular art technique of revolutionary politics) that referenced Mao and the Cultural Revolution. While there were fascinating connections between the Red Guards of San Francisco and Shanghai, commune members in New Mexico and Sichuan, I also knew there were fundamental differences in terms of social practices, subjectivities, and relationships to state and society. Before exploring this further, let me turn to the second example, bringing us to Yunnan Province in the mid-1990s. While at first glance the example appears to be a quite standard case of Western development workers coming to evaluate a Western environmental project in a developing country, on closer inspection it turns out to be much more.

Shifting Winds, Changing Visions: WWF's Project Is Evaluated

International conservation efforts in Yunnan began in 1986, when the World Wildlife Fund started planning a project to conserve what was seen as China's last areas of intact tropical rain forests, only recently "discovered" by Westerners. WWF managed to secure funding from the European Union, and nearly ten years later, in 1995, when the project was being considered for renewal, the EU sent two European project evaluators to Yunnan.[43] After arriving in Kunming, they traveled south, near China's tropical borderlands with Laos, Myanmar, and Vietnam, to a region known as Xishuangbanna (known locally as Banna) to inspect some of WWF's "model villages." Heavy rains and muddy roads hampered their travel, however, and instead they wandered the markets of Banna's main town, Jinghong, where they met with Chinese officials involved in WWF's project.

At the time of implementation (1986), WWF's project plans fit globalized conservation trends well. Tropical rain forests became one of the world's most important ecosystems, often referred to as the "lungs of the earth," vital in producing oxygen and regulating rainfall. As important as they were, however, rain forests were also said to be a "counterfeit paradise:" forests that looked rich and resilient but were in fact incredibly fragile after people had cut trees and cultivated crops, the soils had eroded, and the area had turned into a permanent wasteland (Meggers 1971). Many conservationists during this period portrayed rural farmers' slash-and-burn agriculture as the biggest threat to the rain forest, and some referred to such farmers as "rain forest arsonists." This contributed to an emerging sense among citizens in

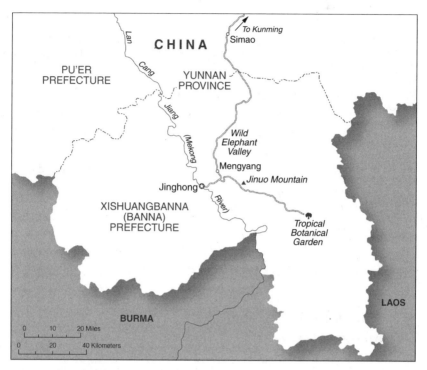

MAP 2. Map of southern Yunnan, showing places described in book.

the Global North that they needed to take dramatic action to save rain forests, a passion that caught people up, motivating people like myself to raise money under the premise that we could buy and protect land.[44] A new international project called "Alternatives to Slash and Burn" advocated agroforestry, a scientific farming method that added trees to hillside fields in order to stabilize slopes and provide the soil with nutrients. Although originally developed at a research center in Kenya as a tool for upland farmers, agroforestry was later seized on by conservation organizations that eagerly began to promote it around the world, particularly in the tropics. One of WWF's main goals in 1986 was to get farmers to stop slash-and-burn agriculture and start agroforestry.

Yet by 1995 globalized conservation trends had changed, and the consultants accused WWF of committing a colossal mistake. The final report stated that local farmers were not destructive but were actually practicing "indigenous shifting cultivation," which the consultants considered the most sophisticated and suitable land use for the tropical forest. Rather than working with the state to restrict local people, the consultants wrote, WWF staff should learn from local people and try to expand locals' rights to land and resources. The EU team offered the following counsel: "As a general rule indigenous shifting cultivation is both more sophisticated and sustainable than many settled monocultural alternatives in the tropics.... There is a considerable amount that WWF and others can learn from [local people about] indigenous agroforestry and biodiversity management" (Newman and Seibert 1995: 49). The consultants advocated that WWF staff should essentially invert its role from teaching to learning. The report was surprising to many staff members, as it challenged the original epistemological basis for the project. It turned out that WWF's field staff, both expatriate and domestic, were hardly stereotypical development agents who disparaged local people; they had a sincere interest in local livelihoods and appreciated the quandaries faced by local farmers. In fact many were worried about the eviction of local people as China became an "environmental state" and promoted increasingly forceful environmental mandates. Therefore some staff saw their efforts to promote agroforestry as a way to help villagers continue farming in their villages, now that much of their landholdings had been seized to create a new nature reserve.

From my perspective, we should not see the consultants' recommendations as one more example of Western knowledge formations flowing one way to the non-West. In fact their report was far more surprising to WWF's expatriate and domestic staff than it was to some Chinese experts. For years the latter had been promoting a community approach that was starting to change the shape, dynamics, and tempo of environmental winds in Yunnan. These experts had written English-language reports that were part of a vast set of efforts to unseat the previous framework, which viewed peasants as ignorant and destructive, in part by using the vocabulary of "indigenous knowledge." The consultants read some of these reports, directly leading the Europeans to "discover" indigenous people in China.

These Chinese experts did not see this community-based approach, which valued local knowledge, as simply emerging from the West, nor did they see it as homegrown. A number of them had gone for advanced training to the Philippines and Thailand, where they encountered radical movements for farmers' rights and indigenous rights.[45] As well, some of the more prominent foreign experts in China who were advocating a community approach and demonstrating participatory methods were not from Europe or North America but from India.[46] Overall these experts were more likely to compare Chinese nature reserves to those in Asia rather than those in the United States or United Kingdom. What is significant, however, is that as Chinese experts debated these concepts in Yunnan, they drew on an older legacy wherein, for more than a century, Chinese intellectuals had both shown an active interest in outside ideas and worked under the assumption that these ideas must be not just directly translated but transformed in order to deal with "Chinese conditions" (*Zhongguo tiaojian;* Liu 1995).

What do these two very different examples tell us? American scholarship typically presumes that the Second Wave women's movement was made in the United States and then exported around the globe. The evaluation of an environmental project in a village in Yunnan is ostensibly about the global coming to the local. Yet in each case there are cross-cutting global and local forces at work. In the first case, China's promotion of liberation and revolution on a global stage helped propel the conditions that gave rise to the feminist and environmental movements—movements that in turn traveled back to China in altered form; in the second case, Chinese experts' studies and publications influenced the way a Western project was evaluated by a European team. The evaluation in turn shaped how WWF began to devise new projects, thus impacting how global environmental groups would carry out their future work.

I want to emphasize that in neither of these cases am I suggesting that we retain the notion of global flows and simply reverse the site of origin, switching Western and non-Western places as the source of these flows. Instead I am arguing that the notion of winds allows us to explore how globalized formations travel back and forth across national boundaries and larger global divisions (east/west, north/south) and how they are transformed along the way. These movements can be difficult to track and do not always comply with predicted causalities, but they show us unexpected connections. Moreover, as much as winds suggest global connections, they also foreground the contours of unique encounters that occur in relation to particular places and histories.

I believe that winds provide a more nuanced and insightful way to think about globalization than do global flows, especially when it comes to understanding how globalized formations and social practices change. Winds encourage us to explore how these formations travel, working in relationship with the terrain they are blowing through. Winds suggest an analytic of transformation, not fixity; a sense of multiplicity, not singularity, that does not start and end within national boundaries or always begin in the West but is made and remade through a thousand engagements. As an example of how these winds work, we can think of physical winds, which change the landscape as they blow—wearing away rock, shaping climate, bringing rain, frost, or desiccation. Winds can be prevailing, like the ocean trade winds that made it possible for Europe to colonize the New World. Winds can also be multiple and vary in intensity, from a gentle wind unlikely to draw notice, to a storm that knocks down forests, destroys river levees, and reroutes water courses. Likewise the environmental winds I discuss are felt in the present and also leave traces of their past in bodies, memories, and physical landscapes— traces that reshape the "the global" itself.

MAKING THE GLOBAL IN YUNNAN PROVINCE

In sum, I extend the Chinese metaphor of winds as a concept for understanding how globalized social formations, in this case environmentalism, travel the world in a process that continually remakes what we think of as the global. The concept of winds acts as a foil to the dominant notions of globalization as flows—those asocial forces assumed to flow from the West by their own accord, dominating and submerging the distinctions of other places under their totality. The notion of flows does not allow us to explore the social consequences of globalizing forces as they affect the changing contours of social institutions and the lives of those ostensibly engulfed by their projects. Instead the concept of winds entails global social formations as something created, maintained, and altered by social relationships and practices.

As a globalized formation, environmentalism (like capitalism) does not happen by itself but exists and travels only due to the diverse, everyday activities that are carried out by a range of people. As the Chinese understand it, one cannot avoid powerful winds; one must interact with them in some way. Without engagement, winds dissipate. As winds blow, they change, not automatically or in ways that are necessarily predictable but through their articulations with specific social landscapes, just as, in a physical landscape, the force and direction of wind is affected by mountains, forests, and valleys. The winds are also reshaping these landscapes iteratively, so that each new wind is blowing through a different (changed) place. As I argued earlier, rather than imagining different social formations, such as feminism or civil rights, to be a Western invention that later flows throughout the world, we should instead think of these formations as winds, which are formed by the travels of and encounters with different people, ideas, stories, and practices. Even if a particular wind began in the West, it would necessarily and inevitably transform as it travels around the world, including movements back to its place of beginning.[47] Thus, in contrast to a notion of monolithic flows, the concept of winds allows insight into the culturally and historically specific social practices that are making and remaking the global.

The following chapters explore Yunnan's changing environmental winds by following the surprising events that transpired during the life course of WWF's project, from its start to its various intended and unintended consequences. They explore the initial negotiations by Chinese officials, the project's implementation in a small rural village, the ways that the presence of wild elephants motivated and threatened environmental efforts, and how environmentalism became increasingly linked to contentious questions of indigeneity in China. To highlight how these winds travel and how people create, engage in, and shape these winds, I employ three concepts that elucidate how global connections are established and transformed: transnational work, the art of engagement, and the making of an indigenous space. Chapters 2, 3, and 4 each explore one of these concepts and the role it plays in making the global, while chapter 5 shows how animals can also get caught up in and contribute to environmental winds.

Chapter 2 examines how the first series of transnational conservation efforts were established in Yunnan Province in the 1980s. These connections across nations were accomplished by the vigilant efforts of a wide range of individuals, from WWF staff to Chinese scientists and officials. I argue that Yunnan did not become part of the global circuits of conservation interest and capital merely because of natural endowments or as a result of Western agency. Rather a small but dedicated group of Chinese intellectuals and in-country staff from WWF carried out what I call "transnational work." These made-in-China efforts involved exploring the province's high mountains and vast forests, identifying animal and plant species, connecting these identifications with the criteria of internationally recognized categories (such as "endangered species"), publishing papers, holding conferences, and creating infrastructure that would attract foreign interest in Yunnan. These efforts have rarely been acknowledged or investigated, yet they were part of the structures by which global environmental winds were harnessed, gathered, and transformed, a level of engagement that is effaced in most models of globalization.

Chapter 3 explores the ways that NGOs, government ministries, Chinese scientists, and local villagers engaged with the new opportunities and difficulties presented by environmental projects, funds, and regulations. I explore how WWF's conservation project worked out on the ground, arguing that villagers neither resisted nor accommodated this project but creatively engaged with project workers within spheres of constrained choice. Villagers experienced the project as the most concrete manifestation of the environmental winds, which had already started to affect their lives. With limited resources, mobility, and status, they tried to work with these winds by negotiating their lives and futures with NGO workers, nature reserve staff, and government officials. They hoped, for example, that NGO workers would offer them new forms of livelihood, better roads, or protection from the predation of wild elephants. Struggles over the project reverberated long after WWF left the site, fostering new social divisions and ways of political engagement, thereby demonstrating that winds are not merely ephemeral but shape social landscapes in ways that continue to be felt.

Chapter 4 examines how some winds create paths for additional, unexpected social transformations. One of the key unintended consequences of transnational environmentalism in China was that it helped precipitate new understandings of indigeneity. By the 1990s global environmental efforts were deeply shaped by the indigenous rights movement. Although Beijing positioned itself as a supporter for indigenous rights in the West, it declared that the concept made no sense in Asia, where everyone was equally indigenous. Nonetheless Chinese public intellectuals grabbed on to and reworked the question of indigeneity, turning it from something seen as foreign and inapplicable to China to a force for critique and social change. The chapter follows a controversy in Yunnan over the issue of "sacred lands," a designation that is attached to a particular social category—indigenous people—and linked to sets of rights. Moreover it describes how Yunnanese scientists started to argue for sacred lands and indigenous knowledge and eventually expanded these arguments to foster a broader sense of environmental justice for all (rural) people. The "smuggling in" of indigeneity by activists already harnessing the environmental winds reveals how the production of new social worlds occurs through the creation of particular pathways, not through submerging flows. The unintended complications resulting from this indigenous turn became a source of tension and great inconvenience for foreign organizations and the Chinese government alike.

Chapter 5 asks, What role do wild animals play in global environmental efforts? Previous chapters examine how particular humans, because of their positions of power or simply their ingenuity in redirecting environmental winds, played a prominent role in "transnational work": attracting and maintaining international interest, funds, people, and projects to Yunnan Province between 1986 and 2011. This chapter shows that people did not do such work entirely by themselves; rather they enlisted the support of many actors—including wild animals—to build networks, foster attraction, and connect China with global conservation organizations. Elephants were already playing a prominent role in reshaping social and natural landscapes, but as participants in these new networks their importance spread. They exerted their own power as forceful beings; unlike the shy and docile panda, elephants continued to seek grain from human settlements, and, after police banned hunting and confiscated nearly 100,000 guns from the region, the elephants became increasingly bold. I relate villagers' stories of elephants' cunning, intelligence, and power as well as their perceptions of the elephants' "war" against crops, homes, and human bodies. Elephant violence created new challenges for managing the increasingly politicized relationships between nature and humans, changing the typical expectation that elephants were only victims of human action.

The conclusion shows how the production of new social worlds is taking place in ways that defy our predictions. As the environmental winds were blowing through Yunnan, a number of relationships and practices came into being, as scientists, farmers, officials, and conservationists worked to foster new worlds. Scientists worked to remake natural landscapes, farmers reached out to officials and development agents, elephants began to carry out war, and the concept of indigeneity was harnessed into new trajectories for rural social justice. Rather than understanding globalization as an all-powerful and monolithic force, the notion of winds—how they are always shaping and being shaped by the social landscape—provides a rich sense of the historical dynamics of social formations in action. It allows us to see how NGOs, rural villagers, and wild elephants are creating new social and natural landscapes that remake the global.

TWO

Fleeting Intersections and Transnational Work

IN THE SPRING OF 2000 I was back in Kunming to carry out interviews with some elderly scientists who were early nature conservationists in 1950s Mao-era China. Five years earlier I had spent a year working at a forestry college there and had become friends with a number of these scientists; I was now eager to catch up with them. When I was there earlier, some made fleeting references to the past that I had not always understood, but in the subsequent years I did much reading on Chinese history and could better appreciate and contextualize their stories. I especially wanted to hear what it had been like in the 1980s, when the environmental winds were just beginning to blow and they were just starting to establish connections with WWF and other organizations. I was also fascinated by how they had become conservationists, even at times of great social and physical threat.

I made my way to a five-story concrete structure, a Soviet, modernist building like those that were beginning to be torn down in the central city. Inside the apartment of Professor Yang Yuanchang I was seated at a card table. To my left a pair of finches sang from within a bamboo cage, while to my right a fluorescent light illuminated a large fish tank. On the balcony wild mushrooms and orchid seedpods were drying on bamboo screens; these natural objects were in stark contrast to the building's cold exterior.

Yang proudly showed me his small study, separated from the rest of his bedroom by a curtain. He pointed out a shelf of old hardback scientific books, most of which were in English and had been obtained in the 1920s and 1930s. When I asked about the books, he told me how in 1937 he had bundled them together when the Japanese invaded China's east coast. Like many of his peers, he had fled across the country and wound up in Kunming. He stayed there and started his scientific career in earnest, getting the opportunity in the 1950s to work with the famous ornithologist Zheng Zuoxin (Cheng Tso-hsin);[1] together they carried out a wide range of work to understand the diversity and range of animals in Yunnan. They worked not only in the lab but also carried out expeditions to remote areas to collect specimens with guns, traps, and nets. They did not collect randomly, but purposely sought unusual specimens that they did not recognize and were almost always assisted by local hunters who escorted them into scientifically unknown mountain ranges and shot many of their specimens. Next they had to preserve the mammals and birds and ship them back to the city, where piles of boxes awaited classification.

All told, Yang studied over seven hundred stuffed bird and mammal specimens, carefully comparing them to scientifically known species, mainly using older books published in English and other European languages. Once, Yang was excited to discover that his supervisor had collected a bird unknown to science. This was, he told me, one of the most exciting days of his life, when he felt the thrill of being the first scientist to know about a new creature for China. He knew that many would be proud but that others might challenge the claim that this was indeed a new species. He worked hard to protect the bird's plumage, known as a voucher specimen, with powdered chemicals in order to ward off insects that threatened to consume his specimen and thereby undermine his claims.

In talking to Professor Yang and his colleagues I gained my first real sense of how much work goes into nature conservation. Maintaining species is not just a matter of leaving them alone; it involves everything from population surveys to instituting new laws and policies and fostering new subjectivities and desires. Yang often led scientific expeditions that sought to determine the health of populations of black-necked cranes, primates, or elephants. The longer I knew him, the more I learned about the often hidden labor of conservation and the ways that transnational connections, conflicts, and collaborations played a role in shaping some of China's dramatic social transformations. Their stories gave me insight into the kinds of labor that I call "transnational work" that create the possibilities for connection. Whereas many scholars portray transnational connections or global flows as seeming to happen almost by themselves, I show that in these cases transnational work was carried out in three steps. First, scientists found examples and labels that mattered to people outside the nation, that possessed some degree of attraction; the idea that China possessed an ecologically viable tropical rain forest was surprising to many conservationists in Europe and North

America in the 1980s. Second, participants looked for points of conversion or traction; in this case, both Chinese and Western conservationists condemned overpopulation and soil erosion. Third, participants elaborated these points and used them to build negotiations and agreements. There were many times when attempts at articulation failed, when conflicting understandings or diverging understandings were never resolved or necessarily acknowledged. Looking at how such connections are made or not can highlight the kinds of contingency involved in these transnational relations and show how deeply all of the participants were affected by the already existing social terrain on which they acted.

Overall the efforts carried out by Professor Yang and his mentor Zheng Zuoxin, among others, turned out to be a crucial aspect of transnational work that helped position Yunnan as China's most ecologically important province and bring greater global attention to China. Together with their colleagues, they created the species lists that showed Yunnan had the greatest biological diversity in China and in much of the world. Although Western scientists had named many of the fauna in this region between the mid-nineteenth century and the early twentieth (a kind of taxonomic gold rush), after the Chinese revolution ended in 1949, almost no Western biologists were allowed to carry out fieldwork in China, with the exception of some Soviets. Thus since 1949 almost all of the species lists were made by Chinese scientists. In the early 1980s they began to published a number of their studies; these reports slowly percolated to scientists in other countries, many of whom were quite curious, especially as data about China were scarce.

Yang also described how he kept hold of his books during difficult times. He declared his avid commitment to conservation until the late 1950s, when those advocating such beliefs were severely prosecuted. Some of his books had covers that were mottled with black mold after being hidden in an underground chest in his old house for almost two decades, during a period when Red Guards searched homes for contraband, including texts in Western languages that would qualify as representing "bourgeois ideology." In the 1980s Yang received an official apology for his suffering under the Cultural Revolution, which was largely blamed on the "excesses of the Gang of Four."[2] He told me that soon after this, he began feeling the impact of the environmental winds, when government agencies asked for his help in conducting surveys to demarcate new nature reserves. It was then that he placed his books, now even more valuable and rare since so many other copies had been destroyed, back on the shelves. He became more intimately involved in transnational conservation in 1986, when he was asked to help guide Prince Philip of England, who came to Yunnan on behalf of WWF.[3]

WWF was the first major international environmental organization to work in China.[4] Ever since WWF chose the giant panda for its symbol in 1961, it had sought to work in China, the only country where pandas live in the wild. The panda was already one of the world's most well-known, well-loved, and charismatic species; WWF's extensive global engagements have only made it more famous.[5] For almost two decades, WWF requested permission to work in China, but its attempts were rejected time after time (Schaller 1993).

In 1979 WWF staff again tried contacting Chinese officials. This was the same year that China had begun "opening up" to capitalist countries. The timing was good for another reason: it followed a recent tragedy during the 1970s when a massive area of bamboo withered, causing many pandas to die from starvation and convincing officials to solicit outside assistance, something they had been loath to do earlier. Thus WWF staff were invited to China, where they spent days negotiating and feasting with Chinese officials. Some staff thought that the officials took advantage of temporal leverage, as 1981 marked WWF's twentieth anniversary, and that a panda project would be a substantial symbolic coup (Schaller 1993). In the end, WWF agreed to commit over a million dollars to panda conservation, probably the biggest funding package for a single wild animal at the time. This was far more than WWF had been planning to spend, but the Chinese drove a hard bargain. Within a year WWF sent the renowned wildlife biologist George Schaller to conduct panda research.[6] After two years WWF established a small office in Hong Kong, which was still under British control. It continued to work solely with the pandas in Sichuan Province until 1986, when officials from Yunnan told WWF staff that their province had intact tropical rain forests, which deserved conservation assistance.

WWF's scientists were skeptical, believing that the area was too far north of the equator to sustain tropical forests, or that, if it once had forests, it was too damaged by rubber and tea plantations. England's Prince Philip was WWF's new president, and he offered to travel to Yunnan to investigate the claim. At the last minute the Chinese officials in charge of arranging the trip contacted Professor Yang when it became clear that the Beijing-trained translators knew little of the natural world.[7]

Professor Yang helped make Prince Philip's trip a momentous event in the history of Yunnan Province and the nation. Beyond fostering China's

Fig.2. HRH Prince Philip resting with Chinese scientists, on ascent of mountains. (Photo.courtesy: Julian Calder/Impact Photos/WWF).

Fig.3. HRH Prince Philip walks through the leach-infested tropical rainforest in Xishuangbanna, Southern China. (Photo.courtesy: Julian Calder/Impact Photos/WWF).

Fig.4. HRH Prince Philip examining the tree panoply in the Xishuangbanna tropical rainforest, Southern China. (Photo.courtesy: Julian Calder/Impact Photos/WWF).

FIGURE 7. Prince Philip with Chinese scientists during a WWF trip to Xishuangbanna, 1986. Photograph by Julian Calder.

pride that it possessed tropical rain forests,[8] Prince Philip's tour initiated new kinds of interactions for Yunnan, fostering a particular set of links with global environmentalism. Professor Yang was a gracious host and guide, building on his decades of work in the region. The prince was enthusiastic about Yang's efforts, the indisputable presence of remaining stands of old-growth rain forest, and the knowledge that wild elephants still roamed. On his recommendation, WWF began its second project in China, which was somewhat different from its work with pandas in Sichuan, often marked by great difficulty, red tape, and mutual suspicion (Schaller 1993). Although this project too was in China, working in Yunnan required engaging a whole different set of officials, experts, villagers, and animal species.

The WWF project was just the beginning of Yunnan's articulations with major international organizations. Beijing was actively reaching out to other global organizations, and during the 1980s it joined the International Monetary Fund, the World Bank, and the Asian Development Bank. The work of Chinese experts attracted the latter two organizations, as well as many others, to start conservation projects in southern Yunnan. Together they devised, funded, and carried out development projects aimed at saving the tropical rain forest by creating new sets of practices and relationships among people, plants, and animals. They aimed to convert locals from destructive slash-and-burn farmers to accomplished agroforesters.[9] Moreover their goal was to inculcate new environmentalist sensibilities and a love of wild nature by connecting urban populations with free-ranging animals, especially wild elephants.

MAKING THE GLOBAL FROM THE GROUND UP: TRANSNATIONAL WORK

What is significant is not that Prince Philip's trip suddenly opened a floodgate for international environmental activities into Yunnan—what some would describe as an inevitable global flow. Nor was it the case that without the prince's trip or the professor's guidance globalized environmentalism would never have come to Yunnan; these winds were already starting to blow. Rather I suggest that this trip initiated a particular set of transnational connections that were actively created and maintained on the ground by the substantial involvement of hundreds of actors, including Prince Philip and Professor Yang. In order to understand these processes, I use the term *transnational work* to signal the diverse kinds of labor that go into making these articulations, not only an initial moment but also ongoing relationships.[10] In so doing, I want to highlight the fact that such connections emerge only through social practices and in relation to particular sociohistorical landscapes—and moreover that any linkage is an accomplishment, and a contingent one at that.

This chapter digs deeper into the transnational work that was necessary for global development projects to gain traction in this particular setting. I suggest that these projects were determined neither by asocial global flows nor by WWF's capacity to impose itself on China. Instead I show that these projects could not have been enacted without the transnational work that was carried out by WWF employees, Chinese scientists, officials, and others. At all levels, the conditions of possibility were deeply shaped by the past and the existing social landscape in which they maneuvered. Various groups involved in hammering out plans searched for places of convergence, areas where differing and emerging interests could intersect.[11] Yet, as the chapter reveals, even after such convergences are created, they can remain tenuous and fleeting and can be quickly dissolved when one group or another changes its orientation.

We have already seen that in the 1980s, men like Professor Yang guided foreigners like Prince Philip and were vital to WWF's initial forays into Yunnan. At first, many of WWF's key interlocutors were men of Yang's generation, already fluent in English and college educated before the Mao era. In order to understand how these Chinese experts worked with WWF staff, let us first take a brief tour of Yang's career, showing how he came to be chosen as WWF's initial guide.

TRANSNATIONAL LINKS BEFORE AND DURING THE MAO ERA

Although many describe the 1980s as the time when the "bamboo curtain" was lifted and China "opened to the world" for the first time, Professor Yang, like many of his peers, was raised in a transnational milieu. Many of the scientists of his generation grew up in relatively wealthy families in east coast cities and were sent to private English-language schools, often Christian. Yang's parents felt that after China abolished the imperial exams in 1911 there was less incentive to have their children receive an orthodox, Confucian education, memorizing ancient classics. Like many of his elite peers in the 1920s and 1930s, Yang was convinced that China's success within the global ecumene would require a new cohort of Chinese scientific experts. This "age of openness" was an exciting time, as Shanghai and other port cities were part of bustling scientific networks and international exchanges. Many Chinese went abroad for training, particularly to Japan, England, the United States, and France; foreign experts were feted in China; and a massive translation effort was under way as Chinese intellectuals and others debated topics such as Darwinism, democracy, and the role of science in China's future.

In 1937 Japan invaded China's east coast, and Yang, like many others, escaped to the interior. There were two major destinations: Chongqing and

Kunming. Chongqing, in Sichuan Province (immediately north of Yunnan),[12] became the interim seat for the ruling Guomindang Nationalist Government under Chiang Kai-shek. Kunming became the interim home for three of China's elite universities,[13] which had relocated first to Hunan Province to escape the Japanese, and Yang followed this intellectual migration. In part it was this war-induced exodus that turned Kunming from a place at the edge of the kingdom, where the emperor sent criminals to exile, into a bustling center of academic talent.

Kunming was a key place of transnational circuits in other ways. It became the end point of the Burma Road, built by the Chinese and used by Americans and British to bring military convoys to support China. Later, after Japan seized the Burma Road, Kunming became the destination of the Flying Tigers, volunteer American pilots who flew over the Himalayas with war supplies from India. In underequipped planes, they fought Japanese bombers that threatened Kunming and Chongqing. After the war ended in 1945 and Mao came to power in 1949, China's elite universities relocated back to their former campuses in the east, but Yang, like many others, stayed in Kunming; they became some of the key figures who would later bring Yunnan into Chinese and global prominence.

In the late 1940s Yang and his fellow scientists traveled to Xishuangbanna (often called Banna) on the same mule caravan routes that had been used for centuries; traders in tea, elephant ivory, and cloth brought goods from present-day Thailand, Laos, Vietnam, and Xishuangbanna north through Kunming and then on to greater China or into Tibet and India. The trip took nearly forty days, and they were quite impressed with the primeval forest, resonant with the harrowing calls of hornbills and other tropical birds and monkeys. In 1956 Yang and others recommended that several sites in Xishuangbanna be designated as a nature reserve (Wu 1965). Their reports traveled to Beijing, where they were well received and approved. During this time, the mid- to late 1950s, government officials approved the creation of a handful of nature reserves, and many conservation-oriented scientists were excited about state support of their work.[14]

Yet after a rapid transition symbolized by the Anti-Rightist Movement,[15] many scientists were labeled class enemies and sent to labor reform camps. Others remained, but during periodic campaigns a number of these scientists were labeled, in the parlance of the time, "old stinking nines" (*lao chou jiu*), one of the nine "black categories" of despised individuals, including landlords and counterrevolutionaries,[16] and many were persecuted by their students and peers. They were accused of wanting to protect the natural world at the expense of peasant livelihoods and national development. As Yang said, "In Chairman Mao's time, it was dangerous to want to protect the environment. Grain was for feeding people, not for feeding birds."

During these turbulent times, despite its relatively small size, Banna was becoming an important political border shaped by important transnational relationships. Banna covers nearly twenty thousand square kilometers (about the size of New Jersey or Israel), yet it is only about one-twentieth of Yunnan's total area, which is somewhat smaller than California and larger than Germany. Banna bordered French Indochina (later Vietnam), where China was sending weapons and supplies to assist Ho Chi Minh's struggles against the French, who were receiving advisors and weapons from the United States. Some of the Chinese soldiers enlisted to help Vietnam were recent veterans, having just fought the Americans in the Korean War. Many were worried that if the French and Americans defeated the North Vietnamese, they might press on into Yunnan (Zhang 1996). As we shall see, the militarization of Banna affected later efforts at nature conservation, and the history of warfare with the United States inflected WWF's negotiations.

Although Banna had been selected as a site of nationally important nature reserves, the Cold War context was having other effects as well: it was turning into a strategic zone for China's key industrial resources. A U.S.-led embargo against China meant that its tropical areas, such as Banna, which cover less than 2 percent of China's land area, were of vital strategic importance for growing rubber. Rubber was a key resource, critical for industrial and domestic uses (tires, shoes, belts for machinery), and the Soviets assisted in designing and building new bridges across the Mekong River to transport troops and rubber. International experts declared that China would not be able to grow rubber successfully in Banna, as it was too far north and too high in elevation (Wong 1975). Yet the spirit of revolutionary science emboldened rubber promoters with confidence and zeal (Shapiro 2001).[17] Thousands of urban youth came from throughout China to open what they saw as Banna's "wastelands" and plant rubber. After some devastating freezes destroyed rubber plantations in the higher uplands in the 1970s, the warmer lower elevations (below nine hundred meters) were prioritized for rubber, the same lowland tropical rain forest area most prized by WWF. One of the designated nature reserves recommended by Yang and others was converted into a state rubber-tree plantation. Even with setbacks such as the freezes, China's rapid success in growing and tapping rubber was a surprise to many in the world (Cheo 2000). It was seemed to show the power of Maoist science to overcome "regressive, bourgeois science" that was tainted by capitalist thinking (Esposito and Lie 1971: 36; Sturgeon and Menzies 2006).

During the 1960s and 1970s, in fact, there was little sense that the environment should be protected from rural citizens. After the Great Leap Forward resulted in a mass famine, state officials quietly imported grain and loudly encouraged Chinese farmers to expand their fields (Fung 1972).[18] Officials did not always try to wipe out shifting agriculture, but as I discuss in more detail later, they encouraged farmers to use shifting methods to increase production of grain and hemp for use by the state (Xu et al. 1999; Mueggler 1998). Grain was a major priority, and rural citizens would be obligated to pay their annual taxes in grain, not cash, for decades to come.

The Reform Era: Changing Views of Science, Farmers, and the Environment

Around the time of China's reform era, starting in 1979, some scientists, including Yang, and officials were part of changing understandings of science and society. Yang described how in 1982 what he called the "rehabilitation winds" meant that China had restored the prestige and value of science and nature conservation. For him and others, rehabilitation meant that university administrators apologized for his suffering during the Cultural Revolution, promoted his rank, and requested his assistance in studying potential nature reserves.

As scientists' privileged position as experts was restored, the notion that villagers could be experts or possess valuable knowledge was diminished. Before the 1980s, state rhetoric praised villagers, who were hailed as inventive, as "model citizens" in building socialism, and as creative practitioners of "Maoist science" (Bray 1986; Schmalzer 2008). Afterward, however, villagers were often described as narrow-minded, reluctant to engage in the new market economy, and incapable of scientific thinking. Terms such as *socialist science* that were popular during the Mao era quickly became tarnished and disappeared, and science was said to need trained experts, not peasants.[19]

Ideas and practices associated with the environment also began to change substantially. Whereas from the 1950s to the 1970s wastelands (*huangdi*) referred to lands, including forests, that were not yet cultivated, by the 1980s the term started to take on a new meaning: places that had been cleared of natural vegetation, farmed, and then abandoned. In my examination of official reports written in the 1980s, I found fewer and fewer uses of *huangdi* to describe land that was not yet cultivated, such as natural forest. At this time, too, one started to see the first statistics that described, with alarm, how little of the natural forest remained and that urged strict conservation (Wang 1979; Lu and Zeng 1981; Wang et al. 1982; Anonymous 1979).

Scientists' concern about the forest paralleled a major transformation in property rights. Much has been written about the massive changes to agricultural lands during the early reform era of the 1980s, when communes were dismantled and distributed to families. Yet another significant change for the lives of millions was the division of forests and grasslands. The state took away huge areas of forest lands that were previously allocated for local subsistence use. Almost two-thirds of China's forestlands had been open for local use, but in the early 1980s state foresters claimed most of this land as national property (Liu 2001; Sturgeon 2005). This meant that, overnight, many villagers' everyday actions became illegal, although for as long as they could remember, they had practiced swidden farming and cut wood in the forests. This tenure shift provided for a massive expansion in nature reserves and other forms of protected forest, which encouraged latent conservationists such as Yang and increased their range of possibilities for promoting nature conservation.

Some of these same Chinese scientists were starting to worry that the continuing expansion of rubber was having negative ecological impacts. In 1979 seventy-seven researchers signed and delivered a petition to Beijing, calling for a halt to rubber planting at the expense of natural forest (Tang et al. 1988: 15). Questioning the expansion of rubber was still a delicate issue, as it had been considered a vital national priority and therefore beyond critique. But after Nixon's visit to China in 1970 and the end of the U.S.-led embargo, China gained easy access to rubber from abroad, and it was no longer a military necessity. At the same time, Chinese meteorologists in Banna began arguing that increasing regional deforestation (including the conversion of rain forests to rubber plantations) reduced the thick fog that had been common during the winter dry season, which was a critical source of moisture for the rubber trees (Zhang and Zhang 1984). The meteorologists argued, in other words, that replacing rain forest with rubber plantations was leading to rubber's decline, as rubber was more likely to go dormant in winter with less moisture (Li et al. 2007). According to Yang, talk of limiting rubber's expansion possibility started to gain traction among officials, especially because it seemed like a problem that might damage the existing rubber plantations themselves. On the other hand, when ornithologists complained that rubber plantations reduced bird habitat (Yang et al. 1985), their reports gained less attention, as there were few advocates for wild birds.

Overall, though, Chinese officials began to identify slash and burn, not rubber, as the major culprit in forest destruction in the 1980s. In the 1970s, officials had lauded the labor of rural peasants who cleared the forest with shifting agriculture, but by the 1980s officials and scientists were starting to describe these same practices as wasteful and antiquated, ones which should be "prohibited and suppressed everywhere" (Yin 2001: 10). In part this changing view was expressed in long-standing stigmas around slash and burn—that it represented a primitive form of agriculture linked to an early stage of human development. For some time, paddy rice was seen in China as the highest stage of agriculture, and slash and burn the lowest stage, which used, it was said, "knives" rather than "plows" (Yin 2001). In the 1980s, for the first time in the People's Republic of China, scientists published studies on slash and burn that used quantitative methods to describe it as ecologically damaging (Yin 2001). Scientists such as Xu Benhan, Yin Yigong, and Chen Zongyi argued that swidden was destroying the forest and must be stopped. Chen declared, "One rarely sees primary forest along the thoroughfares of the area.... If the destruction of forests to open up virgin land is continued in Xishuangbanna, the mountains will be increasingly taken over by grasslands, shrubs, and Eupatorium" (in Yin 2001: 11). This refrain, "grasslands, shrubs, and Eupatorium," became the key symbol of a wasteland.[20] Such studies were bolstered by new research that accused slash and burn of being the major source of soil erosion, which, in turn, ruined sloping fields and clogged dams with silt. The vilification of slash and burn, however, was gradual and incomplete, and many local-level officials eschewed such ecological concerns as they tried to ensure the well-being of farmers in their jurisdiction.[21]

Yang and a group of conservation-minded colleagues traveled to Xishuangbanna in the early 1980s and encountered this notion of farmers' rights, which they saw as a kind of pre-environmental holdover from the Mao era. Although they were confident that they could convince local officials about their plans to carry out stronger measures for protecting forests, they failed. The local officials told them, "As long as the people (*laobaixing*) need to eat, we cannot stop their farming."[22] As Yang put it, beginning in the late 1970s, there was a growing body of research against slash and burn, but this slowly, as he said in English, "percolated" to the officials. According to Yang, higher level officials were the most advanced (the most receptive to global ideas and new federal laws), and the local-level officials were the most concerned about the personal welfare of their subjects.

This history was little known to WWF staff when, after Prince Philip's endorsement, they came to Yunnan in 1987. Older scientists, like Yang, were eager allies of WWF, yet the mere overlap of belief between this older generation of biologists and WWF staff was not enough. Together they faced many difficulties, especially with local officials in Xishuangbanna, as well as higher level officials in Yunnan and Beijing. As mentioned, WWF arrived in Yunnan when there was a growing concern for the environment and when slash and burn was being increasingly vilified by officials and scientists.[23]

It is important to remember that WWF's success had much to do with these recent and dramatic historical transformations in China, as well as forms of deliberate transnational work carried out by its staff, governmental officials, and Chinese scientists. As mentioned earlier, it was critical that regional Chinese officials were changing their own perspectives on appropriate land use before WWF arrived. Although WWF staff and Chinese officials had very different understandings of the past and the current situation, they were able to find areas of shared concern regarding peasant agricultural practices. This provided a fleeting convergence that opened the door to some fruitful collaborations. These conditions were necessary but not sufficient; it took concerted transnational work to create compelling arguments and enlist the necessary networks of support to craft the project and get it funded. They faced numerous difficulties.

Obstacles to Transnational Work

In particular, because it was identified by many in China as "American," WWF and its allies had to overcome Cold War legacies. For decades a number of campaigns called on people to resist the United States, which Mao called China's "most respected enemy" (He 1994).[24] Some provincial officials told me that they had worked hard during the initial meetings in the 1980s to get permission for WWF to carry out its activities in Yunnan, unsure that they could secure a deal. They highlighted three main difficulties. First, they reminded me that such a relationship with foreign organizations was unprecedented. China had a brief honeymoon period with the Soviet Union, which offered much technical and financial support, but always under the condition that China was the "younger brother," in an inferior position. Since the split between China and the Soviets in the 1960s, foreign delegations visiting China typically came as guests and were brought on carefully scripted tours that showcased China's advances in industry, the education system, agriculture, and health care. For decades these foreign guests, often labeled "friends of China," wrote glowing reports about China's rapid achievements, and hosts emphasized presenting a good "face" (*mianzi*) to the world. Now, however, a significant shift occurred in officials' ways of presenting themselves to foreigners: WWF asked officials not to tout their accomplishments but to reveal their social and environmental problems and ask for assistance.

Second, Chinese concerns about sovereignty were heightened because of the particular history of Cold War military tensions. Such tensions were also embodied: a number of Chinese officials and American WWF staff were of the same generation and had fought in opposing armies in Korea or Vietnam. Both foreigners and Chinese harbored mutual suspicions, and modern Chinese history makes frequent reference to the recurring dangers of foreign invasion or intervention. Transnational negotiations were haunted by these middle-aged men's memories of combat, wars in which Chinese soldiers fought directly and indirectly against American soldiers. One WWF staff member was a Korean War veteran, and he told me that during a meeting a provincial official shook his hand almost painfully and said something under his breath that he understood to mean that he too had fought in Korea.

Third, WWF encountered some difficulties because they did not negotiate with a unified state but with several agencies and institutions within a complex hierarchy, some of which competed over access to WWF's international projects, funds, and opportunities to travel abroad. Some agencies tried to monopolize opportunities and benefits for themselves, and this competition led to several WWF staff having difficulties in obtaining their travel visas and other necessary approvals, all of which needed agency permission and assistance. As much as these agencies experienced conflict over how to deal with WWF, such dealings remained marginal to their overall interactions, which mainly concerned domestic affairs. The arrival of WWF and other conservation organizations meant much more to the personal lives of the handful of individuals, especially scientists, who contracted with them for short- and long-term employment.

The Critical Role of a Younger Generation of Scientists

When WWF finally began operating in China, its staff started to notice discrepancies between different generations of Chinese scientists. WWF often first sought information and support from China's older generation of scientists, those in their seventies, such as Yang Yuanchang, Xue Jiru, and their colleagues. Many of them had gone to private English-language primary schools. The middle generation, in their forties and fifties, were often less capable in English, as China's university system had largely ground to a halt during the turbulent Cultural Revolution (1966–76).[25] The youngest generation, those in their thirties, were likely to have greater familiarity with contemporary international concerns and vocabularies. They are the main group I described in the preface, and in the 1990s they were the key group working for international organizations and starting their own domestic NGOs, reaching out to allies and funders, especially in Hong Kong, Japan, Europe, and the United States.

These young experts were not politically powerful in comparison to high-ranking officials, and they were not particularly rich, especially in comparison to wealthy businessmen, but they played a key role in fostering Yunnan's engagements with globalized environmentalism. They were especially important in carrying out the transnational work in the 1990s and later that put this region on the maps used by international conservation organizations and enabled Yunnan to attract and maintain a large number of NGOs. Sometimes one person would work simultaneously for two or three NGOs. These experts were deeply engaged in new theories of conservation biology and trying to understand and carry out work around "indigenous knowledge" and "social forestry" in Chinese conditions.

Younger Scientists Built on the Labor of Older Scientists

Yet it is also important to remember that these younger experts and WWF staff relied on the field and lab work already carried out by the older scientists. This was part of the work that built a foundation of knowledge that could easily translate across nations. In particular they relied on lists built during years of labor by Chinese biologists like Yang and Xue, who had combed the mountains conducting species inventories that evidenced the claim that Yunnan was "China's kingdom of plants and animals."[26] Creating these inventories began again in the 1980s, and this required much knowledge and hard work. Scientists had to know where to lead expeditions and which plants and animals were common and which were rare. They had to track down rare plants and pick and press them, find animals and shoot and stuff them, and then bring these specimens back to the laboratory. The old books, such as the ones that Yang had saved during the Cultural Revolution, were of upmost importance in identifying their specimens, for they were necessary to triangulate findings and plan future collections.[27] Scientists needed to identify each species they found and determine if it was already known to science or a new species. If already known, it might not yet be documented as one of Yunnan's possessions; these totals were built one species at a time. For example, by the early twenty-first century biologists claimed that Yunnan had eighteen thousand species of plants, a statistic that represents an unbelievably vast effort by hundreds or thousands of individuals (Yang et al. 2004). Finding a new species could bring glory to the individual, the institution he or she worked for, the province, and even the nation. The vast majority of this research was published in Chinese, but several of the older scientists worked hard to publish in English, usually in journals based in the United States or Europe, often coauthored with foreign scientists (Li et al. 1982; Wood and Huang 1986; Li and Walker 1986).

The lists Chinese scientists compiled highlighted Yunnan's high levels of biodiversity and, in turn, attracted the attention of prominent conservationists, such as Norman Myers, a British biologist. Myers was making a world survey, trying to determine the most biologically rich places on the planet, especially in terms of endemic species that live only in relatively small areas. In 1986, the same year WWF first went to Yunnan, Myers discovered the publications of three Yunnanese botanists and read their scientific papers, all published after 1982 in English-language journals from the United States and China. Myers corresponded with them in English, and two years later he wrote an influential paper arguing that global conservation efforts should prioritize the world's twenty-five most important ecological "hotspots." Based on the botanists' articles, Myers (1988) declared that there were two hotspots in Yunnan, one of the smallest regions in the world to earn two designations, which also brought Yunnan more worldly attention.

Through the work of many people, Myers's proposal gained traction, and by the late 1990s one of the main organizations to advocate his approach, Conservation International, brought millions of dollars in funding to Yunnan. Thus Yunnanese scientists' efforts were used by a British booster to gain international attention and stimulate more connections, showing how transnational work can occur in diverse and nonlinear ways. These Yunnan scientists did not do their work with the hope of gaining Myers's attention, but written and published in English using scientifically correct and current Latin names, their work traveled more easily than that of their peers who wrote in Chinese.

GLOBAL INTEREST IN THE RAIN FORESTS

As WWF's links to Myers's hotspots shows, convergences between WWF and Chinese officials did not come about just because of internal changes within Xishuangbanna; they also coincided with major changes happening throughout the world during the 1980s. Conservation groups began expanding their previous focus on protecting species, especially "charismatic megafauna" such as tigers, lions, and elephants, and started focusing on conserving habitat (Caro and O'Doherty 1999). One of the key habitats was tropical rain forests. Although previously, rain forests were known as "jungles" or even a "green Hell," by the late 1980s they became valuable (Slater 2003). Rain forests were frequently in the news and quickly became emblematic of a growing fascination with and concern about the environment (Hecht and Cockburn 1990). David Sobel (1989: 19) describes the emerging sensibility that pervaded this time, and how concern for tropical rain forests shaped the lives of young children in places like the United States:

> Just as ethnobotanists are descending into tropical forests in search of new plants for medicinal uses, environmental educators, teachers, and parents are descending into second and third grade classrooms to teach them about the rainforest. From Brattleboro, Vermont, to Berkeley, California, school children are learning about tapirs, poison arrow frogs, and biodiversity. They hear the story of the murder of activist Chico Mendes and watch videos about the plight of indigenous forest peoples displaced by logging and exploration for oil. They learn that between the end of morning recess and the beginning of lunch, more than 10,000 acres of rainforest will be cut down, making way for fast food "hamburgable" cattle.

Sobel offers us a sense of the pervasive structures of feeling that were going on elsewhere around the globe, animating people like myself, who in the late 1980s were deeply concerned about rain forests despite never having been to one. This was also a time when there was a growing sense of responsibility, which linked consumption in North America and Europe to ecological damage in tropical forests. I joined protests against companies like Burger King (accused of buying "rainforest beef," as Sobel referred to it) and Home Depot (accused of buying unsustainable tropical hardwoods), although I later realized these issues were more complicated than they first appeared. During the late 1980s the tropics became a more substantial part of everyday politics, linking consumption to social movements and a sense of globalized consequences. By this time there was frequent reference to the expression "The personal is political," which, as mentioned earlier, was a phrase coined in 1970 by Carol Hanisch, a Mao-inspired feminist organizer.

During this time dominant ecological theories supported the idea that rain forests were incredibly fragile and particularly susceptible to human disturbance, a "counterfeit paradise" wherein an apparently robust fecundity masks a true vulnerability (Meggers 1971). According to these theories, the rain forest was constantly under threat because there was only a thin layer of soil supporting the vegetation. If this layer of soil was removed, the rain forest might never recover, turning instead into wastelands. As understood by rain forest biologists, wastelands were permanently damaged lands that would neither return to their original condition nor be agriculturally productive. The fear that rain forests were becoming wastelands, a term that resonated quite strongly with the Chinese notion of *huangdi,* motivated many international projects, including WWF's work in China.[28] At the time, many scientists in China and elsewhere viewed slash-and-burn agriculture as the major cause behind increasing wastelands, and therefore as a key threat to rain forests (Myers 1984).

DEPICTING TROPICAL NATURES

One of WWF's first priorities, and one of their first examples of transnational work after the prince's 1986 visit, was to obtain funding from the European Union and permission from a range of Chinese officials to carry out their project. Specifically their effort to frame and legitimize their mission in Xishuangbanna rested on assertions that there were three major problems: overpopulation, land degradation, and slash-and-burn farming. These claims had enough correspondence with Chinese notions to easily gain some initial traction and support, but further developing these connections required concerted transnational work by individuals with differently situated plans, intentions, and goals.

I discovered these efforts in part through interviews with some of the expatriate and Chinese workers involved in this project and through archival research. The archival work and interviews privilege the perspective of two WWF staff members, John MacKinnon and Jack Bentley, in part because I had good access to WWF's archives in Beijing, whereas the staff at the Xishuangbanna Nature Reserve were not forthcoming with their archives.

Dr. John MacKinnon came to Xishuangbanna after having arrived in China in 1987 to work on WWF's panda project in Sichuan Province, a project that revealed some underlying tensions between expatriate conservationists and Chinese government officials (Schaller 1993). A British biologist, MacKinnon is a prolific author, professional photographer, and committed environmentalist. Although his doctoral studies at Oxford focused on wild orangutans in Borneo and Sumatra, he later expanded his expertise to other mammals throughout Asia (MacKinnon 1974). WWF hired him as a senior advisor for their panda projects. MacKinnon is energetic, strong-willed, and often unafraid of controversy. In 1993 he became well-known in China for his reaction to a request by the Chinese government for $100 million in foreign funds for panda conservation. This was a tremendous jump: in 1979 WWF's $1 million contribution seemed massive. MacKinnon told reporters, "It's a hostage situation. The Chinese are pointing a gun to the head of the panda and saying, 'If you want to keep it [from going extinct], fund it. Otherwise we're going to let it go'" (in Sun 1993). By the time MacKinnon began work on WWF's Xishuangbanna project proposal, some Chinese officials told me, he was known as confrontational and "troublesome." Conservation, in other words, was never apart from domestic and international politics but part of shaping these politics.

MacKinnon's report on Xishuangbanna, like many environmental documents, characterized the landscape as both ecologically rich and "under threat" (MacKinnon et al. 1996). In his view, the greatest threats were the twin forces of population pressure and land degradation. In turn, these threats were made manifest through slash-and-burn farming. Some of his colleagues referred to these farmers as "arsonists of the world's heritage." At the time, satellite images of tropical forests burning in Brazil and elsewhere were increasingly available to conservationists and the public, and this created much fear that rural people in developing countries were burning the "lungs of the earth" (Soroos 1998: 317). Landmark best sellers like Paul Ehrlich's *The Population Bomb* in 1968 helped instill a deep fear of overpopulation and reinforced fears that exponentially increasing populations were destroying majestic rain forests merely to raise corn, rice, and cattle (Myers 1984).[29]

POPULATION PRESSURE

Although many organizations in the West were describing China's birth-planning system as draconian (Aird 1990), a number of Western conservationists in the 1980s, such as MacKinnon, regarded it as an effective way to curtail overpopulation. On this count, conservationists tended to laud China as an environmental model (Davoll 1988). MacKinnon's claim that "population pressure" was the key reason behind land degradation required almost no work to legitimize in the eyes of WWF's funders such as the EU or Chinese officials, even if the former didn't condone China's methods of population control. The idea that global overpopulation was a major cause of environmental problems was already seen as established fact. China in particular was and continues to be one of the most frequently cited examples of the detrimental ecological and social effects of overpopulation; as Susan Greenhalgh (2003) points out, the academic and popular press almost never questions the claim that China is overpopulated. In fact it would have drawn much skepticism if MacKinnon had argued that population pressure in China was not the major factor in land degradation.[30]

Although international agencies and Chinese officials were likely to agree with MacKinnon's assertions about the problem of population pressure, his particular explanation had difficulty gaining support within China. Conservationists often explained high birth rates as the outcome of rational, yet unfortunate, choices made by families to create more hands to share the farm work. This was often portrayed as a rational yet environmentally tragic "vicious cycle" of expanding populations and decreasing yields.[31] In China, on the other hand, explanations for higher birth rates could not rest on a similar frame of logic. The state's insistence that urban families should rationally desire one child and rural families two children was hegemonic. There was no space in China for suggesting that the desire to have larger families, with three or more children, made any rational sense. Families wanting more than two children were not viewed as rational but as representing a feudal consciousness (*fengjian yishi*) that should be changed through threat of penalties or "thought work" (*sixiang gongzuo*).

MacKinnon suggested that certain groups, especially non-Han ethnic minorities, were largely to blame, as they had high birth rates and "destructive lifestyle habits" such as hunting. He wrote that the ethnic minorities' burgeoning population was attributable to officials' reticence to strictly police the minorities, as well as minorities' lack of foresight and inability to voluntarily limit their own birth rate. Although MacKinnon claimed that birth rates remained high in rural Banna, local officials would not verify this situation, and it would have been difficult, if not impossible, to confirm. Even if it were true, local officials who were in charge of enforcing birth quotas feared persecution from higher levels of government if they admitted such facts. In other words, while population pressure was a suitable explanation when discussing environmental problems at the national level, local officials were leery of being accused of allowing higher birth rates. Thus MacKinnon's claims were simultaneously supported, challenged, and ignored by Chinese officials at different tiers of the governmental hierarchy, from national to provincial to local levels.

MacKinnon also declared that ethnic minority "habits" such as hunting must be stopped (MacKinnon et al. 1996). His perceptions and language are not idiosyncratic but closely echo biases found in the majority of Chinese texts about ethnic minorities (Gladney 1994; Litzinger 2000; Schein 2000). It is very likely that MacKinnon heard such biases from his Chinese peers, as I myself heard many times during conversations in the 1990s and later. He also could have imbibed such biases in the very few English translations that were available at the time, such as Tang Xiyang's *Living Treasures: An Odyssey through China's Extraordinary Nature Reserves*, published in 1987. Tang, a Beijing-based reporter, traveled to Banna and visited an upland ethnic minority village, which he described as destructive and wasteful, using Marxist terminology about their "level of production."[32] Tang (1987: 63) wrote, "Their primitive level of production and low standard of living have caused much damage to the forests. Each year these people burn down large sections of the forest as part of their system of cultivation. They use axes rather than saws to cut down trees to build houses, often abandoning large logs they cannot chop after cutting off only a few branches." In this example, and there are many more, Tang castigates their actions as backward, inefficient, and destructive. They seem to exhibit no concern for the ramifications of their actions, which are deemed wasteful and damaging. MacKinnon's work, in turn, reinforced such existing notions of destructive ethnic minorities for his Anglophone readers and funders.

It is important that we not be presentist in adopting a negative dismissal of such portrayals, for MacKinnon largely followed existing notions in China that valorized the lifestyle of lowland, valley-dwelling groups and vilified the upland groups that lived on mountainsides. He produced a movie for WWF that juxtaposed lowland and upland groups (1991). In it he described the lowland Dai ethnic group (historically the majority group in the region, with the most wealth and political power) as good environmental citizens who raised paddy rice, cultivated diversity-rich home gardens, grew their own fuel wood in groves, and maintained "sacred forests." He described upland ethnic minority groups (rarely specified by particular names), on the other hand, as practicing superstitions and destructive methods of slash and burn.[33] MacKinnon's movie drew on English-language publications by ethnobotanists like Pei Shengji, who valorized the Dai. (Years later Pei's student, Xu Jianchu, was one of the first to powerfully challenge negative views about upland groups by documenting their indigenous knowledge and their sustainable methods of shifting agriculture and raising rattan vines, as explored in chapter 4.) Overall MacKinnon's position on minority groups fit smoothly within China's existing ethnic hierarchies.

Moreover, at the time, MacKinnon's views on conservation also coincided with dominant ideologies held by many Western biologists, especially those of his generation. These included strongly held beliefs that conservationists needed to strengthen the state's role in enforcing birth control, implementing natural resource laws, and permanently removing rural groups from ecologically valuable areas (Igoe 2004). Thus MacKinnon's description of threats to the rain forest resonated both within China and abroad, helping to enable transnational work that created successful articulations.

Such beliefs were in part motivated by the widespread notion among conservationists that human actions, such as hunting, farming, and gathering wild plants, were intrinsically harmful to "wild nature." Beyond the narrow worlds of academic anthropology and ethnobotany, almost no studies before the late 1980s described such human actions in environmentally positive or even neutral terms. For MacKinnon, there was no possibility that Xishuangbanna's upland ethnic minorities (*shaoshu minzu*) might count as indigenous peoples who possessed valuable knowledge and globally recognized indigenous rights. While the lowland Dai might be regarded as ecologically virtuous, in the late 1980s no one was attempting to position them as an indigenous people, as recognized by international law. Chinese experts had not yet tried to make indigenous people into a category in China that would have social and political impact.

Instead the reigning view of Western conservationists was to see rural people in China and elsewhere as peasants, trapped in a vicious circle of ecological destruction, fueled by their ignorance and an inability to understand the damage they caused. Problems lay almost entirely with the peasants themselves.[34] Solutions were to be provided by outsiders who could instruct peasants in rational and scientific methods of family planning and farming. This emphasis on science as a solution meshed with interventions by the Chinese state, which promoted a particular suite of activities in rural areas, such as "modern" pig raising, biogas generation, and aquaculture—practices deeply embedded within scientific frameworks that were similar to those of conservation organizations (Schmalzer 2002). In other words, both the state and conservationists were involved in projects to transform "backward" rural dwellers into scientific citizens, or at least ones more productive and less ecologically harmful.[35] Much of the environmental writing of the time was inflected by an urgent, alarmist tone and fears of imminent ecological collapse that demanded immediate intervention in order to forestall disaster.

WRITING ABOUT THE DANGERS OF SLASH AND BURN

Diverse forms of transnational work did not end with MacKinnon's initial proposal to secure project funds from the EU. Instead more participants were hired to create a detailed plan of action. The two main fieldworkers for the project were Jack Bentley, WWF's principal consultant, and Yang Bilun, the main Chinese coordinator. Bentley is an American with development experience in the Philippines. In 1987 he met MacKinnon during a birdwatching trip in Xishuangbanna. Several years later MacKinnon hired Bentley and Yang to carry out the project. Bentley wrote most of the reports and carried out trainings. Yang coordinated a vast number of logistical and bureaucratic details (most of which Bentley was never aware of), drove the WWF vehicle, and translated between English and Chinese. Yang's and Bentley's perceptions differed quite substantially, and, as the men were located in quite different institutional networks, they engaged with different audiences and issues.

I first met Yang Bilun in 1995, and our relationship has strengthened over the years. When I returned for fieldwork in 1999, he persevered in the difficult task of obtaining permission for me and my family to live in Xiao Long and accompanied me there for the first ten days. My only face-to-face interactions with Bentley were in 1995, but we have corresponded over the years. He left a collection of reports at the WWF-China headquarters that assisted me greatly in understanding his perceptions of the events in Xiao Long.

Yang Bilun was born on New China's first National Day, October 1, 1949. By the time I met him and his family in 1995, they had for decades suffered much at the hands of state agents and fellow citizens. During his time in elementary school, his parents' status was a well-known fact, and he faced daily humiliations and estrangement from his peers. At school he was at times forced to wear a dunce cap or sit for hours on three-legged chairs. College students pasted a large character poster over the doorway to the Yangs' apartment. The poster, which covered the top half of the entrance, denounced the Yangs as class enemies. The Yangs left the poster in place for years, ducking under it as it sagged, too fearful to remove it.

Like millions of other formally educated youth of his time, in 1968 Yang Bilun was sent away as part of a political reform campaign, "down to the countryside and up to the mountains." He was assigned to work in a Jingpo minority village, near southwestern Yunnan's Myanmar border, seen as a place of much hardship and poverty. His identity as a class enemy followed him, and his freedom of movement was notably less than that of his urban peers, who were often distressed by rural lifestyles and did their best to get positions in towns. As he described it, Yang Bilun made the best of the situation and helped build a village school, where he taught for years, creating strong relationships with many of his Jingpo students. Unlike almost all other outsiders, he learned how to speak Jingpo, an unwritten language at the time, by inventing a system of transcription based on the Russian orthography he had learned in middle school. His marginality and his time spent in a poor rural minority area led him to develop what is in China a heterodox appreciation of culture and cultural difference.[36]

When the Cultural Revolution ended in 1976, Yang Bilun and the other educated youth were allowed to return to their urban homes. Back in Kunming, he was assigned to drive a forklift in a factory. In 1979, as part of China's "opening up and reform period" and for the first time in many years, the government held a national college entrance exam, and against the competition of over a decade's worth of applicants, Yang did very well. Consequently the state selected him to study rubber production at the Tropical Forest Experimental Station on Hainan Island off China's southeast coast. Converting tropical forests to rubber groves was considered of vast importance and an expression of patriotic duty. Students were often reminded that the rubber trees were important state property and constituted the "means of production." The trees were a long-term investment, and fear of permanent damage from improper tapping meant that students were punished for slight mistakes when trying to coax rubber from the trees. During the off-season the students trained by peeling the skin of oranges in thin layers, using rubber-tapping knives.

At school Yang Bilun also studied English, a decision that was critical to his later work with NGOs. After China split with the Soviet Union, Russian was no longer taught and English eventually became part of the official curriculum. Ironically, Yang's parents' difficult past suddenly became a valuable resource. He was always encouraged to study hard, and his early exposure to English, learned through songs that were sung in secret with his parents, gave him a distinct advantage over his peers. There was almost no contact with native English speakers until the late 1980s or 1990s in many parts of China. According to Yang, "We were the start of the 'silent English' generation, where you can read, and hear somewhat, but not speak." His own English-speaking ability distinguished him from his peers, but this ability was of little use for a number of years. After graduation he was sent to Southwest Forestry College, where his parents were on the faculty and where he taught students about rubber management and traveled throughout Yunnan, researching insects that, like rubber, were valued as a state strategic resource, yielding industrial tannins and shellac. He continued in this vein until the late 1980s, when WWF's John MacKinnon contacted Yang Bilun's father, the prominent biologist Yang Yuanchang, requesting him to organize a birding expedition in Xishuangbanna. Professor Yang recommended his son to lead the trip. Yang Bilun's connections with his Hainan classmates who had been sent to Xishuangbanna's rubber institutes helped immensely in planning the trip to a part of China where officials normally allowed foreigners to stay only within a tightly prescribed tourist zone. It was during this trip that Jack Bentley became involved in the initial plans for the WWF project and got to know MacKinnon and Yang. The bird-watching trip created the social context for these men to form their initial friendships and then working relationships with the agroforestry project.

Agroforestry is a system of raising certain kinds of trees together with annual crops and is intended to address soil degradation in developing countries (Nair 1996). Agroforestry was often designed for nonmechanized farming on steep slopes, and it aimed to reduce villagers' dependence on chemical fertilizers. It often used tree species that fertilized surrounding annual crops with nutrient-rich leaf litter and nitrogen released from tree roots (using nitrogen-fixing trees). In part, it was intended to replace slash-and-burn agriculture, which was seen as a rising threat to the world's tropical forests.

Bentley's report articulated a global concern around slash and burn and provided local data and examples of soil erosion. But he wasn't compelled to verify his general narrative, because the belief that slash and burn was the leading threat to tropical forests and the cause of impoverishment of rural peoples was already so hegemonic among development and conservation NGOs.

Some of these assumptions actually made little sense in China, but it appears that there was little input from Chinese scientists.[37] As well, the proposal was sent to the EU, where those who evaluated it were likely unfamiliar with China, as there were almost no detailed studies of China's rural human-environmental relationships in any European languages, let alone accounts that differed from the dominant perspective. Bentley (1990), for example, stated that "shifting agriculture with less fallow time between declining yields on severely eroded mountainsides continues to cause the need for ever larger families." This theory likely originated in other places in the Global South, but was it applicable to China? Even elsewhere, was soil erosion the main reason behind increased family size? Did this really happen in Yunnan, where, in my own experience, most rural minority families follow guidelines and have just two children?[38]

Although some scholars argue that powerful organizations such as the World Bank create a monopoly on authoritative knowledge in places like Laos or Egypt (Goldman 2001; Mitchell 1988), this was not the case in Yunnan. Bentley did not rely on WWF scientists to produce facts for his report. Even though WWF was one of the world's largest conservation organizations and carried out extensive studies on panda habitat in Sichuan, its scientists did not do any ecological research in Yunnan for the project. Instead Bentley assembled facts recently produced by Chinese scientists and combined them in an attempt to "localize" global development narratives.

Like many foreigners who examine China's conservation literature but cannot read the language, Bentley tended to ignore sources written in Chinese, instead focusing on the relatively few texts already translated into English.[39] He relied heavily on bilingual Chinese scientists who carried out field studies and published their reports in English in the 1980s. He did not use any reports from the 1970s or earlier. This is because prior to the onset of the environmental winds, there were almost no studies that blamed villagers' actions for eroding soils or damaging natural forest. In the 1980s, however, Chinese experts began conducting experiments that were already premised on the idea that farmers' practices were a key problem; their reports, with quantitative statistics and written in English, created the conditions of possibility for Bentley to assert his claims and use local facts as evidence.

Such facts emphasized the state of looming disaster. Indeed their thesis that Yunnan was a catastrophic case of land degradation seemed clear:

> 39% of Yunnan suffers from a serious erosion problem and 10% has a very serious problem. (Bentley 1990: 2)
>
> One act of swidden loses 149 years of soil accumulation. (Bentley 1990: 2)
>
> [Wasteland] now makes up almost half of the land in [Xishuangbanna] prefecture. (Cheung and MacKinnon 1991: 23)

Like "population pressure," these facts of land degradation were so self-evident to conservationists during the 1980s that few thought to challenge these statistics. This was not unique to China, for there is little evidence that scientists in Asia, India, or Africa questioned these percentages or the definition of categories such as "serious erosion" and "wasteland." There was a wide-scale pattern of blaming peasant farmers for ecological ills. By the late 1990s many scholars started to challenge such claims and present counter-narratives arguing that rural farmers were not to blame for desertification, deforestation, and soil erosion (Leach and Mearns 1996; Showers 2005; Turner and Taylor 2003), but at the time of Bentley's and MacKinnon's reports, these statistics bolstered existing notions and heightened fears of an impending crisis.

On the other hand, as WWF and other international NGOs carried out their work in Yunnan, they also fostered the social production of this knowledge. At least since the late 1980s negative characterizations of slash and burn were becoming more commonplace, described as wasteful, destructive, and threatening. Regional authorities were starting to promote laws against slash and burn, but these laws had not yet gained much social force or momentum. Although WWF did not directly carry out its own studies, it did fund the salaries and equipment used in some of these Chinese erosion studies. In particular it bought GIS equipment and vehicles, helped nature reserve officials and scientists attend international conferences and workshops dedicated to stopping slash and burn, and directly sponsored a number of development workers in Yunnan. This involvement further legitimized and fortified the authority of a stance against slash-and-burn agriculture that was growing among Chinese officials.

WWF's largess was particularly striking at a time when it was quite difficult to gain access to such goods as vehicles and fax machines. In the early 1990s, when an entire university, with thousands of faculty, staff, and students, shared access to just three vehicles, one professor obtained a car for work and personal use with WWF's panda symbol stenciled on the door. As well, invitations to regional conferences in Asia, facilitated through WWF, could provide the initial impetus for a scientist to obtain a passport, which was the most difficult hurdle that experts faced in traveling abroad. Once acquired, however, a passport could open up many possibilities.[40] Groups like WWF could also help scientists obtain the funds required to travel—not just money but foreign currency needed for international travel. At the time, Chinese currency was not officially exchangeable abroad, so Chinese had to obtain foreign currency within their own country. The favored currency was U.S. dollars, which banks exchanged only when accompanied by forms with official permission. Alternatively some scientists could use their own *guanxi* networks, their social connections to obtain U.S. money or in some cases get permission for a passport, but this was often difficult. Thus, as an organization, WWF provided critical cultural and material capital for those who worked with them, as well as the means by which to strengthen the transnational networks these scientists wished to participate in more actively.[41]

TRANSNATIONAL WORK AND MAKING CONVERGENCES

Although they maintained differing perspectives and goals, staff from WWF, the Xishuangbanna Nature Reserve Bureau, and natural scientists from several institutes worked together to create a feasible plan of action that would include representatives from each organization working out zones of overlap. One of the key areas of overlap in WWF's project was reducing soil erosion. This choice had numerous advantages and helped to create a convergence in several ways. First, campaigns against soil erosion had a historical resonance in China as many Mao-era projects aimed to conserve and improve agricultural landscapes. Millions of villagers built terraced fields, planted trees, and dug canals, all with the aims of enhancing agricultural yields and conserving water and soil. Most people over fifty, whether living in rural villages or in cities, whether scientists or officials, had participated in these campaigns, and this helped WWF's project form a recognizable bridge into less familiar concerns about protecting "wild nature," a concept with little historical traction.

Second, as some scholars have shown, NGOs and other organizations had to work within the parameters of acceptability (Ferguson 1994). The choice of working with farmers was expedient because it presented a politically appealing approach. There were debates among WWF staff as to their course of action, and in some cases the range of possible interventions was narrowed by their desire to not confront the Chinese state. Seeing farmers as the problem did not threaten the state and, in fact, meshed with postreform narratives that peasants were in need of training and improvement. Conversely, while MacKinnon privately condemned the spread of rubber plantations into natural forests, other WWF staff decided that they should not complain about this, as they did not dare openly confront the state-run rubber industry. MacKinnon also endorsed campaigns to increase the enforcement of birth control and confiscate guns from villagers, but his colleagues feared that such suggestions from a foreigner would offend Chinese officials, and they wisely asked him not to mention this in his project proposals. Teaching peasants scientific forms of agriculture, however, in no way threatened state sovereignty and sat well with state visions and projects.

Thus WWF staff did not have the ability or desire to impose their vision singlehandedly. Instead they carried out many forms of maneuvering that tried to establish places of correspondence between themselves and Chinese officials and to avoid sites of conflict. I suggest that such maneuvering is key to the ways that international NGOs operate in China and elsewhere. Such actions—what I call transnational work—play a critical role in opening up and trying to maintain a space of collaboration. It is not necessarily true that either side had full or even much awareness of the other's position, but there was enough congruency that some alliances were created, at least temporarily. As we shall see in chapter 4, such convergences began to fall apart by the mid-1990s.

Third, and more important, the focus on erosion was advantageous because it was part of an internationally prominent development narrative (Roe 1991). In order to attract NGO attention, staff at the Xishuangbanna Nature Reserve and Kunming-based scientists needed to describe ecological problems and solutions that were recognizable, such as tropical deforestation, soil erosion, and agroforestry. Likewise NGOs working in China needed to use dominant global narratives in order to obtain support from major funders like the EU. The description of erosion and its causes in WWF's project proposal meshed with the received wisdom held by members of other international organizations. Just as important, fighting erosion was considered desirable by Chinese government officials. Chinese scientists in particular had little to gain and much to lose by questioning this position, as almost all of them were based at state-run institutes and colleges and feared reproach if their research challenged the official position. Of course, not all of the scientists supported these dominant narratives, but those who did not often found themselves with fewer publications and grants and less job security.[42]

Fourth, soil erosion offered a specific and practical focus for the project. NGOs attracted funding only when they both described problems and offered solutions. In a number of ways, these strategies converged enough that WWF's plans were seen as logical, politically feasible, and important. With funding in hand, narratives in place, and support from the Nature Reserve Bureau and Ministry of Forestry, Bentley began to carry out his project in Xiao Long, one of the villages chosen by staff at the nature reserve for WWF. At this point Bentley's energies switched from negotiating in boardrooms with Chinese foresters and writing project proposals for EU administrators to working with villagers, trying to convince them to give up slash and burn and commit themselves to science through agroforestry.

SUMMARY

WWF staff were able to carry out their project because of the ways they, together with Chinese scientists and experts, were able to carry out transnational work, mobilizing and working across differences. They did not do all of the work themselves, however, for they stood on the shoulders of many, like Norman Myers, who made the importance and plight of tropical rain forests a key global issue. Looking first at the efforts of WWF, people like

George Schaller and John MacKinnon helped WWF beat all of the other international organizations to be the first to work with China's pandas and rain forest, so they became the initial global brokers to mediate between international funders and the Chinese government. By convincing the EU committee that Xishuangbanna was an important site, WWF sought and gained funding at a time when the EU was giving out massive amounts of international development aid, and reducing tropical deforestation was one of its top priorities. Its problems were seen as knowable and hence solvable. When MacKinnon and Bentley arrived, they relied on a set of established and intermeshed global narratives about population pressure, the environment, tropical forests, slash-and-burn agriculture, and peasants. Their reports both drew on and entrenched these beliefs. At the time, some of the WWF staff's views overlapped with those of Chinese officials. On the other hand, many points of divergence remained, just a few of which I have elaborated.[43] Regardless, these points of confluence, in part created through extensive transnational work, afforded some advantages for each group. Local officials used WWF's support to gain domestic prestige with Beijing, leveraging an icon of international science and funding to reinforce their own perspectives and plans. WWF, in turn, relied on Chinese scientific reports to buttress their claims and needed permits and permissions from officials to carry out their projects.

There were several major areas of congruence between the approach of officials and that of international conservation organizations. First, both groups perceived a serious crisis in a landscape now understood as a critically important ecosystem. Second, this crisis was mostly blamed on local people, whose practices were perceived as ignorant and unscientific. Third, both groups looked for solutions based in science. In many ways, WWF's claims converged with that of the state, at least for the time being. This convergence, accomplished through a wide variety of transnational work by a multitude of differently located agents, reveals the complex and shifting social dynamics behind what we might otherwise interpret as a global flow, which either diffuses effortlessly or is imposed with an iron fist.

The notion of transnational work shows that these international conservation groups that are carrying out projects in China do not just get to impose their own plans, and neither do they always work in tandem with the state. Rather at present there are a multiplicity of agents that attempt to carry out transnational work and create new spaces of collaboration. The concept of transnational work does not apply only to international projects focusing on the environment, health, or rural development. It can also be used to analyze the proliferating connections that enable international capital investments. For example, pioneering anthropologists Lisa Rofel and Sylvia Yanagisako are building on a legacy of insightful research on the power dynamics on factory floors (Ong 1987; Rofel 1992; Pun 2005) to track the quickly changing relationships between Chinese factories and Italian fashion firms, and how they rely heavily on translators' creative work. One could also look at the ongoing ways in which entrepreneurs negotiate changing World Trade Organization regulations, activists work to link domestic and international human rights NGOs into fruitful partnerships (Merry 2006; Keck and Sikkink 1998), and marketers finesse complex transnational supply chains (Tsing 2005).

As we shall see in chapter 3, Bentley's attempts to carry out WWF's agroforestry project met with some resistance by villagers but also with surprising amounts of interest and engagement. Such dynamics were strongly influenced by the role of the Chinese state and by the particular sets of claims made by representatives of a transnational organization. In turn, each group was fractured and found itself moving in different directions, and this momentary convergence shaped later outcomes in unforeseen ways.

These conjunctures did not just arise, but were made, in part through finding areas of traction that linked people across national boundaries in shared interests and concerns, such as tropical rain forests, soil erosion, and overpopulation. When Chinese experts and officials began to view soil erosion as a major problem and carried out field studies and surveys that gave this vision an empirical validity, this was one of the conjunctures created by many people. Therefore these conjunctures are not the grounds upon which transnational work is performed but are actually part of the work itself. Participants from many sides worked to elaborate these points, and build projects and agreements. Transnational work emphasizes how what may feel like powerful flows that come from nowhere are actually the product of the cumulative actions of many people, not all with the same intended effect but that combine and change in powerful ways.

THREE

The Art of Engagement

I ORIGINALLY CAME TO XIAO LONG ("Little Dragon" village, pronounced *she-yao lawng*) to understand how WWF's project, from 1988 to 1995, had worked out in one of its "model villages." Based on my readings in anthropology and cultural geography, I expected to see nature conservation efforts as another form of development, which, as a matter of course, imposed hardships. These readings also led me to anticipate that villagers had a strong desire to maintain their own autonomy and resisted WWF's attempts to impose new ways of life. In many accounts from many different locales, local people oppose global efforts at control. But residents quickly disabused me of my expectations. I learned that they did not so much automatically eschew such outside efforts, whether from international NGOs or national agencies, as they in fact tried to reach out and solicit new social connections. I now understand their efforts as a way of practicing the art of engagement, which is neither a complete rejection nor an encompassing embrace. Although I was originally focused on WWF, I found that it played a relatively minor role in shaping village life compared to a long history of state actions. I also learned that I could make sense of this process only within the broader sweep of time. This is because villagers' attitudes toward WWF staff were deeply shaped by their everyday encounters with officials and other powerful outsiders, in play long before and long after WWF left. Thus the ideas and practices they brought into their engagement with the environmental winds and WWF specifically were shaped by previous interactions. The arts that villagers practiced were not entirely new ones.

It took me some time to appreciate this, for in 2000 Xiao Long seemed like a quintessential anthropological field site, seemingly beyond the reach of most global forces. Around two hundred people lived in forty-two ancient-looking houses built with thick posts and beams, perched above a small valley lined with fields of paddy rice. Tea bushes grew on gently sloping fields to the east, and the whole village was surrounded by tall tropical forest, within which lived China's last remaining wild elephants. There were few nearby villages, and Xiao Long's residents were officially classified as an ethnic minority, specifically as Jinuo.[1] I found that no one who still lived here owned a car or motorcycle, and they rarely went to market, producing almost all of their own rice, vegetables, and meat. When WWF's project director Jack Bentley came in 1988, it looked even more "traditional," as there was no electricity yet; people built water-powered rice pounders that they placed in the creeks to pound paddy rice, and no one owned a TV. At that time no one had a two-wheeled tractor (*shou fu tuolaji*), and paddies were plowed by water buffalo.

By the time I arrived, electrical lines had been installed, and most families had TVs, DVD players, and stacks of Hollywood movies (sometimes watched in homes with split bamboo walls and a grass thatch roof). Almost all the plowing animals were gone. Furthermore I soon discovered that the residents of Xiao Long were not an organic community at the edge of state rule; rather their peripatetic movement and village landscape were intimately bound up with state campaigns and transnational military affairs. In fact, to describe themselves they often invoked a Mao-era song about being from *si hai jiu zhou* (four seas and nine states), an old Chinese term meaning far-flung places.

As Xiao Long residents spoke a local dialect, it took me time to grasp the ways their speech differed from the Mandarin dialect I was taught, and I was grateful for people's efforts to try out their best Mandarin on me. Most families were in a similar economic and living position and by the 1980s were starting to tear down their homes with bamboo walls and grass thatched roofs and replace them with wooden post-and-beam houses and tile roofs. Many of them planted cash crops of *sha ren* (*Ammonum villosum*, or Chinese coriander, a valuable medicinal plant) in the forest but ate almost exclusively the food they grew themselves: three meals a day of rice topped with a dish made of meat sauce, hot peppers, or greens, cooked or raw. Chickens largely lived their own lives, although chicks were protected from snakes at night and fed small bits of broken rice. Eggs were laid in hidden spots around the house and seldom eaten, but boiled chicken was the luxury food par excellence to be served to visiting cadres and foreign visitors. Pigs were more demanding; almost every family gathered pig food from field and forest,

FIGURE 8. Slicing wild banana stalks for pig food, 2001. Xishuangbanna, Yunnan. Photograph by Michael Hathaway.

finely slicing the stalk of wild banana or pounding taro tubers with a large mortar and pestle and cooking the pig's meals every day. Once a year, around the Chinese New Year, most families would slaughter one pig, which would be rendered into a wide array of sausages and preserved meats, to be carefully parsed out over the year, just as the year's rice harvest influenced daily portions.

There was relatively little outside movement, and only a handful of people sought and found permanent paid work. One family moved to town to sell fruit, but later returned when it came time for their grown son to find a village wife; they didn't want him to marry a town girl, who might be loose with money and morals and would not be hardworking. All told, I heard of four young men who left to work as a forestry police officer, a truck driver, a health officer, and a teacher. Residents took pride in these men and made efforts to keep them close to village life; sometimes they could help facilitate urban connections. I heard that family members would cunningly use the young men's health plans, as villagers had no insurance and a costly hospital visit could bring "poverty for two generations." On the other hand, people who had grown up in Xiao Long but moved to the city quickly adopted the smug attitude of many urbanites, viewing themselves as superior. I cringed when one man, the health officer, visited his brother's house and loudly talked to me about the backward thinking and poor hygiene of his brother's family, pointing out several shortcomings: "Look, they soak the rice in water and then feed that nutrition to the pigs—their kids are skinny, but their pigs are fat, *aiya*." The officer aimed for my complicity, as I was a doctoral student from America, assumed to have an "advanced scientific outlook."

Only one resident, a freelance trucker, had attained an independent cash-based living that allowed him to buy a house in town, about an hour's drive away. Years later he was still paying back all of the family and friends who had loaned him money to buy an old truck, a blue, dented *Dong Feng* (East Wind). He was widely admired for his pluck and success, and when people first mentioned his name to me, they often followed it with a term suggesting his daring: *danzi da* (literally, that he had a big gallbladder, the center of bravery in Chinese idioms, comparable to references to the testicles in English). Some said they were jealous of his life in town, and he would periodically bring back cases of beer or bags of baked goods, like cakes, that they could not afford. They jokingly called him "liberated" (*jiefang*) from the daily toil of a farmer's life but were surprised one day when he showed up with his wife and son, a young man who dropped out of school and joined a small-town gang. He set up his wife and son in his old house and brought them occasional supplies, which included a cook stove and a monthly propane bottle, while every other family cooked over firewood. His son, after months of alienated resentment toward what he called a boring village lifestyle, eventually embraced the more macho side of village life. Yet he still struggled with plowing paddy fields using a walking tractor, carrying huge loads of corn back from swidden fields, and spending all day in the forest, trying to keep the weeds at bay in his sha ren patch. He sought out one capable male role model in particular, who also chafed at village life, pining for his teenage years living with his brother in town, where he spent a lot of time chasing girls, drinking, and playing guitar. Yet he also took the young small-time gang member under his wing, glad that his skills, which seemed to him easily acquired and beyond notice, were held in such high regard.

A few young adults in Xiao Long, all male, had graduated from high school, which was a noted accomplishment. School, which had been free for years, was now costly, and testing into high school was competitive. They were often no match for the town-educated kids, whose parents helped them every night with homework and spoiled them with daily treats and nutritious meals. Instead Xiao Long kids attended a one-room school in a small town that was about a three-hour walk away, where the teacher was often unmotivated and hoping to move to a bigger town. Starting at age five or six, boys and girls would board at the public school all week, traveling home on Saturday only to return Sunday. Their parents periodically delivered a sack of rice for them, which the teacher cooked; they knew other places were even more "backward," where children boiled their own pots of rice. Higher education was considered an unworthy investment for many families, and relatively few were willing to pay what they saw as exorbitant fees for high school, especially for daughters, who would typically marry out of the village. Those young men who did finish high school were well aware that their families sacrificed to enable their education. One of them fulfilled his duty by working in the poorly compensated position of village accountant (*kuaiji*), managing village-based expenses and proposing development plans.

The best-known story about the perils of education concerned another young man, Wu Ling, a potentially prodigal son who graduated from high school and finished a two-year associate's degree in physical education. When Wu failed to find paid work and returned to the village, this reinforced existing skepticism toward the value of education. He lived in a kind of social limbo, walking around in a shiny blue tracksuit and chatting with his age-mates, performing a few chores in a lackadaisical manner, weeding the corn fields and feeding chickens. His soft, easily blistered hands and his defeated manner were frequent topics of conversation. He knew little about farming, and it seemed somehow inappropriate to ask him to do dirty or strenuous jobs. Some saw Wu's fate as a sign that society was undergoing a serious change—before, every college graduate had been assigned a stable lifetime position, but now the "iron rice bowl was broken."

Of the young women who left Xiao Long, none had made it through high school, but a number found work in regional restaurants. The work was often lighter than at home, but the hours were long and the pay was low. They left together in groups of two or three and rented a crowded room, sometimes provided as part of their job. These stints were often temporary, as there were plenty of young women willing to work for little pay. This work was an adventure but also sometimes humbling, as other workers claimed not to understand their dialect or made fun of their unfashionable clothes or hairstyles. Some longed to travel to Guangdong, where it was said they could work in factories and save money, but so far none had dared the long train ride, which took at least three full days. They sometimes assured themselves that such work was taken by girls more desperate than themselves, like those from Sichuan, which was crowded and poor. When they returned home, they helped their mothers in largely female-designated tasks like gathering wild herbs for pigs and picking tea. I learned to pick tea, as it was in a public space and an easy way for me to talk to women of my own generation. Even though my picking skills were poor, I could still contribute to a day's work. Talking with the young women and their mothers, I heard that their travels to town were curtailed by rumors about the dangers of the sex trade. No one from Xiao Long had become involved in the sex trade, I was told, but it was alluring, with its promise of more freedom, adventure, and high pay. This was also the time when China finally began to admit its own HIV/AIDS epidemic, and tales of rampant heroin addiction were widespread. Banna is on the northern edge of the historic Golden Triangle, and is a notorious site in China for transporting heroin. No one in the village was an addict, but many had kin or friends who were.

By and large, almost every family had their own land, which mainly provided adequately for the family's subsistence. Many attempted some entrepreneurship, such as raising fish in ponds or selling passion fruit or dried bamboo shoots. These efforts were often short-lived, as the market in passion fruit crashed and finding wild orchids to sell quickly became too difficult. One of the few activities that remained popular and profitable was selling sha ren, which earned a steady income. As the plant could be dried and stored for a long time, it was easier to deal with than goods like fresh fish that were difficult to transport.

One of the few who did not grow his own rice was a man, now alienated from his neighbors, who was also danzi da but who angered his peers rather than earned their respect. He conducted secret negotiations with officers in the army platoon that had stayed in Xiao Long in the 1970s to create the first tea plantations and gained rights to harvest a relatively large area. He built a rudimentary tea-processing plant, using old machinery that was heated by wood. He built a small, poorly maintained house for his workers and several outsider families. These families dealt only with their boss; they were never invited to meetings, weddings, or funerals, and even their children never played with village children, nor did they attend school. I soon learned to keep my distance from the tea boss, for as much as I was curious, I was warned not to be too friendly with him.

This chapter explores the art of engagement, a formulation I use both in homage to and in contradistinction to James Scott's formulation and elaboration of "the arts of resistance" in his books *Domination and the Arts of*

FIGURE 9. Bamboo and grass houses in Xiao Long, 2001. Xishuangbanna, Yunnan. Photograph by Michael Hathaway.

Resistance: Hidden Transcripts (1990) and *Weapons of the Weak: Everyday Forms of Peasant Resistance* (1985). Scott and many other scholars have shown how dominated groups resist and subvert hegemonic control and attempt to keep governing forces at bay through often subtle acts, including evasion and foot dragging.[2] My fieldwork in rural Xiao Long village, where WWF had carried out conservation projects, however, showed me that the broad range of rural residents' behavior was not well described by the term "resistance."

In fact many villagers exerted a good deal of effort trying to connect with and attract powerful outsiders, including those associated with WWF's project; others picked and chose what they wanted from the project, while some simply did their best to ignore it altogether. All of these villagers were in a somewhat precarious position vis-à-vis WWF and the state: they were geographically isolated and economically poor, earning an income of under 1,000 yuan a year (approximately U.S.$120). Moreover insofar as they lived on the edge of what became a highly valued nature reserve, they were especially affected by the environmental winds that were blowing across China (and the world). These winds ushered in a whole new sensibility about the environment; along with a series of pressures and threats, the winds also carried new opportunities. Chinese citizens struggling to live well despite often quickly shifting and powerful winds were keen to anticipate and negotiate these rapid changes, with the aim of building better futures for themselves, their families, and sometimes their village. How these threats and opportunities impacted their lives depended, in part, on how they practiced the art of engagement.

The art of engagement in rural China was seen as a variety of skills enacted with various levels of competence. They changed over time and were carried out toward differing goals. Art, even in the singular, involves a variety of skills. These skills overlap with but also differ from what scholars of China call *guanxi xue* (the art or practice of cultivating social connections), which uses gifts, favors, and banquets to foster social relationships, indebtedness, and exchange (Yang 1994; Yan 1996; Kipnis 1997).[3] I suggest that this phenomenon is not unique to China, nor is it new. Whereas many studies of global encounters are premised on a clash between the imposition of international power and the resistance of local peoples, this chapter argues otherwise. The art of engagement includes what we think of as resistance, but it is much more, because it emphasizes actions in building new relations, deflecting potentially harmful plans, and transforming latent possibilities. The chapter shows how villagers conceived of and engaged with powerful outsiders in a way that was quite different from my expectations of resistance; it then takes a closer look at some of the academic critiques of resistance that have been useful in pushing me to think about the art of engagement. The remainder of the chapter is an ethnographic study of how globalized environmentalism came to the village of Xiao Long and how the art of engagement was practiced not just by villagers but also by WWF representatives and officials, and it exposes the ways they maneuvered in relationship to others rather than imposed their plans. Through this ethnography, I explore what it might mean to shift our focus from resistance models to engagement models of global encounters.

FROM RESISTANCE TO THE ART OF ENGAGEMENT

I first heard about Xiao Long while living in Kunming in 1995. WWF had just ended a seven-year project (1988–95) in Xishuangbanna, and Xiao Long was one of its four "model villages." I had a number of meetings with Jack Bentley, WWF's American project manager, and Yang Bilun, the Chinese project assistant. Bentley had just returned from an awards ceremony in Beijing to recognize their efforts. I continued to correspond with both of them over the years, and I also discovered a lot about the project in WWF's archives. In 2000 I traveled to Xiao Long for the first time, along with my wife and three-year-old son, to carry out long-term fieldwork. From the city of Jinghong the road followed the curving valleys of smaller rivers that emptied into the Mekong, a wide, muddy, and slow river. We passed dozens of brick hamlets built in the 1960s and 1970s as part of state-run rubber plantations. Where the side of the road allowed, there were occasional fruit stands and restaurants, often raised above the ground on posts. They were fashioned from split bamboo and roofed with thatch and corrugated asbestos panels. We pulled off onto a dirt road and traveled to the end after negotiating deep puddles and steep hills. During the summer monsoon season, residents quipped that a four-wheel-drive car was useless in the thick and slippery mud, even a "four-person" car was insufficient (meaning one person to drive and at least three to push). The village was surrounded by tropical rain forest that had been recently demarcated as a nature reserve. I knew that people in Xiao Long had been intimately connected to the Xishuangbanna Nature Reserve Bureau since the mid-1980s.

After WWF's project came to a close in 1995 and their subsequent proposal was not able to secure funding, they left Banna.[4] Even without WWF, however, the power of the Xishuangbanna Nature Reserve Bureau continued to grow, and Xiao Long remained in what some called the Bureau's domain and what some called its "kingdom" (*wangguo*). This domain was a place targeted for monitoring and frequented by forest guards, and it continued to attract new projects carried out by development organizations such as the World Bank's Global Environmental Facility and Germany's Gesellschaft für Technische Zusammenarbeit (GTZ).[5] I was able to stay at Xiao Long after getting permission from the Bureau's leader. He had been a student of one of my older colleagues at the forestry college, who "pulled connections" (*la guanxi*) on my behalf. In turn, I affiliated myself with the Bureau and helped them with translation and with editing project documents for the World Bank and GTZ. I talked with many of the employees, from upper management to field officers, meeting them for meals at their homes and in restaurants. Even though they believed that all of the villages near the reserve were within their territory, the reserve also overlapped other realms of authority, and I had to visit township officials to pay respects and register with the police station.

I heard then that Xiao Long residents, like hundreds of other villagers in the reserve's kingdom, were now struggling with the new and often strict forms of nature conservation that deeply impacted their hunting, swidden agriculture, collection of firewood and medicinal plants, and dealings with wild animals such as elephants—now under the protection of the state—that threatened their fields.

When I arrived in Xiao Long in 2000, I fully expected to see an example of globalization wherein WWF and the reserve staff imposed coercive conservation and villagers resisted their efforts. The papers I had read by anthropologists and geographers in the 1990s were full of examples of how globalized conservation efforts caused hardships for rural people in the Global South.[6] Such studies often described conservation as a form of neocolonialism, a set of practices that unfairly alienated local people from their own lands. Indeed in some places conservation programs expanded processes of dispossession that had been carried out for over a century by colonial and national policies.[7] The stakes for environmentalism can be quite high; in some places, strict conservation laws have resulted in the malnourishment and even the death of rural people denied access to land for farming or gathering edible and medicinal plants (Harper 2002; Turnbull 1972); in other places, subsistence hunting was deemed poaching, and hunters faced heavily armed park rangers with "shoot to kill" orders (Peluso 1993). These consequences of globalized conservation, particularly as they take place in the rural Global South, are part of the underbelly of international conservation efforts (Brockington and Igoe 2006).

Not surprisingly, then, ethnographic studies of conservation often highlight resistance, such as rural people burning down state-claimed forests or carrying out subterfuge to maintain the long-established practices of hunting, cutting timber, and collecting medicinal plants (Peluso 1994). In fact these writings on conservation are part of an older and broader literature premised on the notion that subaltern subjects resist domination, whether in the form of forestry departments, colonial forces, local elites, or postcolonial states (Comaroff 1985; Guha 1990; Peluso 1994). When I came to Xiao Long, I didn't expect that residents would actively protest against the state—I thought the police would quickly stop such attempts—but I did expect to see resistance and a desire for autonomy.

And there was a certain amount of truth to these scholarly observations: Xiao Long residents were well aware that the environmental winds had brought significant changes to their lives and chafed under many of them. Guards now came to inspect the reserve to make sure villagers did not carry out slash-and-burn agriculture, and I heard that a few villagers were fined and jailed, even though the reserve boundaries were often difficult to discern and were only occasionally marked with small concrete posts that sometimes moved or disappeared. More recently the police had confiscated guns and cracked down on hunters, thereby changing the act of hunting from an activity of social camaraderie, especially among young men, to a solitary pursuit, carried out with an intense fear of getting apprehended. The World Bank had just given funding to the Xishuangbanna Nature Reserve Bureau; at the bureau offices I saw a film showing a crack squad of ex-soldiers patrolling against poachers, supported by World Bank funding.

Xiao Long residents talked about environmentalism as a hardship that brought new restrictions into their lives, a description that resonated with those heard elsewhere in the world. Although many in Africa see conservation as a familiar form of neocolonialism with a long history (Adams and McShane 1992), in Xiao Long people often described conservation as an aberration, a betrayal of what they knew about the People's Republic of China. It had been a country that aimed at the improvement of rural people's lives by helping them convert wastelands into agricultural lands and that promoted development through the provision of health care, education, and technical support. Some villagers said that by the late 1990s, things had changed: "The government now cares more about elephants than us."

Yet despite these complaints and the hardships that Xiao Long's residents suffered, to describe their actions as "resistance" would not do them justice. Rather they struggled to engage with the environmental winds, gauging where new opportunities seemed to exist and how they might need to change their practices to reduce the chances of being fined or worse. Residents said that by the 1980s their connections to officials began to wane, and they reacted quickly to potentially fleeting opportunities. All visitors came unannounced, as there was no telephone line and cell phone signals didn't enter deep valleys between town and their village. During the summer monsoons, for months at a time villagers could be fairly certain that officials would not visit. Villagers walked in and out on forest footpaths but claimed, "Cadres [government officials] don't walk anymore; if they can't drive here, they won't come." When residents heard the whine of a vehicle climbing up the last hill, they would strain to interpret its sound. I found that even children had a sophisticated knowledge and taxonomy of vehicles, classifying cadres and guards into different ranks according to the type and value of their car. A visit by officials could signal a new policy with potentially serious implications. Forest guards arrived more often and might surprise villagers hunting along the road or expanding their swidden fields, potentially onto reserve lands. When directly confronted by such an infraction, these guards might feel obligated to act, with potentially devastating implications not only for the individual but the family as well. I heard stories of families that fell into poverty when a husband was jailed, even for a few years, as wives and children struggled with less income and labor. I felt as if I understood villagers' sense of threat, yet it took time for me to learn the various ways that they reached out to officials in the hope of making relationships rather than evading contact or directly challenging the new order of things.

More often than not, and in sharp contrast to my expectations, local residents said that the best way to deal with potential threats was not by keeping officials at bay but by inviting them into their village to share in meals and festivities, to dance, and to establish kinship ties. An open invitation to township officials rarely worked, but sometimes one could attract state representatives by personally inviting them to housewarming celebrations, weddings, or even a youth-day performance. One winter during the wedding season, a number of large, shiny Mitsubishi SUVs were parked in Xiao Long, where almost no one could yet dream of owning even the cheapest car. They were not driven by wealthy friends or relatives but by officers at influential state work units like the Public Security Bureau, which exercised strong police powers. Such officers were seen as more important than those who were merely wealthy. The groom's family bragged for weeks that they were able to attract six SUVs from the prefectural capital of Jinghong—each one a public declaration of the family's importance and a testament to their strong guanxi network.

Equally significant, these attempts to seduce outsiders quite often failed. During the youth-day performance, a village-wide show with rehearsed songs and dances, I noticed Zhang Bing sitting in the front row, his shoulders hunched in deep dismay. In his forties but looking older, Zhang was chain-smoking cheap local cigarettes. Often disheveled, he was known for his eagerness to improve his social status and his long streak of bad luck. The person next to me whispered that Zhang had convinced everyone that three local officials would attend the performance. Yet to Zhang's right, three empty seats cast a pall on the entire show. To Zhang those empty seats represented his weak guanxi; to others they represented the growing insignificance of their village. That night an older man, somewhat drunk on strong homemade rice alcohol, reminisced about their old leader, Lao De, who was in charge when WWF showed up. "Lao De had a gift," he said, "he could put an end to long-time village feuds, and cadres loved to talk to him [*chui niu*, literally 'blow cow,' which might best be translated as 'shooting the shit']." It was this old man who told me that certain people had real skill in engaging others, especially outsiders, who often regarded themselves as superior. He also said, however, that it was increasingly difficult for rural villagers, backward mountain folks like himself, to talk with cadres or entrepreneurs from town, as the gap in their lives and sensibilities seemed to be getting bigger: "Lao De was really skilled, but times have changed."

I was struck by the fact that in 2000, when villagers talked about WWF's project, few, if any, wished that WWF had never arrived. They were not sure why WWF had left, and some hoped for another chance—that another development project would come their way. Some worried that WWF had left because villagers had failed to live up to its expectations. Others suggested, both boasting and with a sense of impotence, that their elephants were too smart for WWF, as the animals foiled its attempts to fence off paddy fields.[8] It was not just their stories that suggested responses other than resistance; their active efforts to reach out often deviated from what I had expected.

Before I further explore the complex forms of selective engagement that I came to understand, I want to take a brief detour through some academic literature on resistance. Although a number of scholars are beginning to express dissatisfaction with the idea that resistance is a simple reaction and now supplement resistance with terms such as *creative* or *critical* (Hoy 2004; Sparke 2008; O'Hearn 2009), resistance remains a dominant model for interpreting global encounters.

THE ALLURE AND PITFALLS OF RESISTANCE

Ever since James Scott's initial elaboration, in 1985, of "everyday forms of resistance" as the "weapons of the weak," his ideas have inspired a great deal of productive and interesting work. Scott's books (1985, 1990, 1998, 2009) explore what Guha (1997) has called "dominance without hegemony," offering new insights for political scientists, historians, social scientists, literary theorists, and others who wish to understand relations of rule and power, and especially how these manifest in the lives of the "weak," often poor and rural subalterns. Although the concept of resistance has been most actively used to describe situations in the Global South, including influential scholarship like India's Subaltern Studies Collective, it has also been used in the Global North.[9] Indeed the concept of resistance has changed the way many historical accounts have been written. Before Scott, scholars often failed to notice diffuse, everyday acts of resistance and instead focused on organized, collective protest and revolutionary change. In the absence of such forms of protest, historical accounts tended to be narratives of domination, which downplayed subaltern agency. Subalterns were portrayed as victims of such forces as modernity, patriarchy, capitalism, colonialism, or globalization. Yet endeavoring to reverse this structure, resistance theories tended to presuppose that subalterns always aimed to keep such forces at bay, to maintain their own autonomy.

Since that time, resistance theories have been subject to a number of criticisms based on their dichotomous framework, reactive view of agency, and romantic portrayal of progressive politics. Resistance studies tend to create dualistic narrative structures that are morally Manichaean, in other words that have (only) two sides: a virtuous resisting local and a villainous dominating outsider, whether in the form of local elites, states, or global forces (Bernstein 1990).[10] Critics of mainstream resistance models argue that subaltern agency is often presented as largely predictable, as reaction more than action (Jefferess 2008; Ortner 1995). Critics claim that scholars often romanticize resistance, eschew the diversity and specificity of subaltern politics themselves, and rarely explain why some subaltern groups resist and others accommodate (Abu-Lughod 1990; Ortner 1995; Brosius 1997; Mitchell 1990). In fact anthropologist Charles Piot (1999: 198) declares that critiques of resistance are an "academic cottage industry."

Despite these criticisms and points, resistance remains a dominant frame for interpreting subalterns' relations with global forces.[11] Why is this? I suggest that the concept of resistance appeals to scholars because it implies that attempts at hegemony are rarely absolute, and it offers hope that the oppressed understand their situation and work against it. On the one hand, there is something like resistance in the world, because attempts at dominance and hegemony are never fully accomplished or stable.

On the other hand, I suggest that our understandings of resistance tend toward analytically impoverished accounts. Most versions of resistance conceive of it in terms of two groups, dominators and resisters, caught in a kind of zero-sum game. Perhaps more disturbing is the inherently conservative side of resistance: it is typically seen as a desire for subaltern world-maintaining. As cultural studies scholar David Jefferess (2008: 11) argues, resistance is typically seen as a "struggle *against* something rather than *for* something." I fully support his suggestion that we should valorize the more dynamic aspects of resistance, "as acts, processes or values that perform a politics of transformation" (7). For example, Jefferess uses this transformative notion of resistance to explain how Gandhi and many other Indians resisted colonial British rule in the 1940s; their actions did not aim to keep colonialists at bay but to *build* an independent, postcolonial India. Such a view would push us beyond accounts of a clash between two groups with fixed positions and would help to invigorate the idea of resistance, allowing us to see what is created through such engagements.

A further step in rehabilitating our ideas about resistance is to challenge its romantic assumptions, along with its implicit link to progressive politics. Saba Mahmood offers an insightful discussion of resistance in her book *The Politics of Piety* (2005).[12] Mahmood argues that resistance needs to be understood in relationship to agency. She asserts that scholars have often ignored how mainstream Anglophone notions of agency are linked to a particular Judeo-Christian vision of the self and progressive action. She is skeptical that one can determine "a universal category of acts—such as those of resistance—outside of the ethical and political conditions within which such acts acquire their particular meaning" (9). In contrast to many who see resistance as acts of freedom, Mahmood moves beyond the question of resistance per se to look at agency and, more provocatively, the role of desire in motivating actions. She asks how desires are produced, and how we explain desires that do not seem to accord with assumptions of progressive politics (e.g., that women strive to be free from relations of subordination [10]).

Although an invigorated notion of resistance and an attention to desire is a good first step, this still does not necessarily offer insights into how two (or more) groups, riven by internal differences and divergent motivations, engage in ways that simply cannot be reduced to resistance or desires. To do this, I suggest we cast a wider net, cultivating an attentiveness to divergent strategies and actions, and explore how social engagements are themselves generative of changing social formations.

I had assumed that villagers would wish for freedom from governmental intervention; in fact I discovered they were often more interested in greater entanglement.[13] Throughout the 1980s they sensed that many aspects of state involvement were fading from their lives. This led, on the one hand, to the fear that they and their village would simply be forgotten. On the other hand, with growing interest in nature conservation, they were concerned that their livelihood was being increasingly restricted, even to the point where they became potential environmental criminals. Given this, villagers often greatly desired to engage with officials, the nature reserve staff, and others.

DESIRE (AND BEYOND) AND THE ART OF ENGAGEMENT

My fieldwork taught me that resistance models account for a very small part of engagement and that people were far less motivated by attempts to keep powerful forces at bay than by the desire to gain connections and cultivate relationships beyond the village—to deflect potential threats but also create new opportunities. When villagers reached out to officials, inviting them to housewarming parties, or when they tried to establish fictive kinship relations, as I once saw happen between a former hunter and a forest guard (when the guard became the "dry father," or *gan die,* something like a godfather relationship to the hunter's child), they were performing the art of engagement. Villagers have limited opportunities to meet higher-up officials and few resources with which to forge relationships, and many feel self-conscious that, unlike their urban relations, they do not have the oratorical skills, nice clothing, or money to do such things as host a large meal at a fancy restaurant (often done in cities to foster relationships with people of greater social standing). Given this, villagers conducted the art of engagement using whatever was at hand.

One resource for villagers was the use of wild animals. Ironically, as a growing number of environmental laws gave more protection to some wild animals, residents in Xiao Long more frequently used them to conduct the art of engagement. Restaurants were more cautious about selling wild meat, as a number of them had been heavily fined after being raided by police. This meant that tourists, wealthy entrepreneurs, and officials, the main consumers of wild meat at restaurants, found it increasingly difficult to acquire, and therefore it became even more valuable. As a result, rural residents turned to gifting wild animals to foster social relations, sometimes with the same officials who could persecute them. One night Xiao Long residents actually managed to lure some township-level officials to attend a house-christening party. At the party I became aware of my own status as an object of attraction, as I was compelled to make toasts and chat with the officials. Later that night one of the men (with me in tow) was tempted away from the party with the promise of roasted wild birds. The official had mentioned that, with the new conservation rules, these were now too difficult to buy in town, and one of the well-known hunters quickly ran wooden skewers through a large bowl full of small sparrow-like birds and roasted them over the coals. I could not refuse. I nibbled around their bony bodies and watched the officials' stiff demeanor soften as he and the hunter reminisced about how much more common it had been to eat wild birds in the past. Decades ago they, and many other little boys, had caught such birds with improvised bamboo traps, pieces of grain, and strings, tying them by the leg to a stick and selling them to neighbors.

In the spring I watched two villagers catch a bamboo rat (*zhu shu*; *Rhizomys sinensis*), pinning it to the ground with an overhand hoe. These rats lived in the forest, creating deep underground burrows and eating the bamboo roots. The villagers gently wrapped the rat in a shirt to keep it from biting, placed it into a cardboard beer box with small holes for air, then smuggled it into the police compound in Jinghong, where it became a gift for an officer. This was not just a random officer but someone that they knew loved to eat bamboo rats and who would relish the opportunity to cultivate his own social networks with other state officials through shared feasting on an increasingly rare culinary treat. The rats were not rare in places with abundant bamboo, but they were difficult to catch and, with the recent laws, much harder to find for sale. Villagers used these animals to create and maintain good relations with officials, hoping that such relations might help them avoid arrest in the future or gain compensation from the Nature Reserve Bureau when elephants or other wild animals ate their crops.

As with any art, some were more skilled at engaging than others. Some villagers failed to speak well, especially with people of higher social status; some failed to offer the right gifts at the right time and place. I heard about big mistakes, whereby the recipient of a villager's gift regarded it as a bribe and became angry at the insult. But villagers also said that some of their leaders had great skill in attracting powerful outsiders, such as township-level officials, tea bosses, and other entrepreneurs who brought wealth and possibilities.[14] In fact Xiao Long, like many villages with some collective savings, had a small fund that could be used to repay villagers for feting outsiders on behalf of the village—to cover the cost of a chicken or a bottle of alcohol, for example. At times this fund was drawn upon to host spontaneous visits by powerful outsiders, but a small group also developed a list of plans for potential village development projects, all of which would need much external funding and support, such as building a paved road, re-opening a school for the youngest children, or erecting a "culture station" (*wenhua zhan,* a small building stocked with books and magazines). They also used this fund to send small delegations to Simao to search for "tea bosses" (*cha laoban*) who offered the best prices for their tea or to Jinghong to negotiate with the Forestry Bureau in attempts to gain back some of their land that had been turned into a nature reserve, or at least obtain compensation. All of these examples reveal that part of the art of engagement is not a form of tradition-maintaining resistance but of forward-looking action, working to build the type of land- and socialscapes that would be able to attract future projects.

In exploring how these dynamics worked out on the ground, I argue that this is not a case of two groups at war, wherein powerful global actors succeeded or failed in imposing repressive regimes on local villagers. Instead I show how several groups creatively engaged with each other within shifting spheres of constrained choice. For example, rather than assuming that WWF was all-powerful, we can look more closely at how its own representative, Jack Bentley, had to practice the art of engagement, albeit with a very different understanding of what was at stake. Bentley navigated with different players a complex set of personal desires and professional relationships that were often difficult and in conflict, but as is typical in such situations, his power to impose plans was compromised (T. Li 2007). Villagers, on the other hand, did not react as a homogeneous collective, as resistance models predict; some welcomed certain aspects of the project and tried to reject others. Sometimes they did so quietly, and at other times they directly challenged Bentley's perceptions and hopes. Important third and fourth parties included government officials, whose approval was necessary to allow the project to continue, and nature reserve staff, who were often eager to build alliances with WWF but were not always fully supportive of Bentley's objectives. Reserve staff and officials wanted the project to be a success but were concerned that creating too much villager resentment could engender difficulties in the years to come. The project itself changed substantially as it was worked out through these various engagements.

Whereas chapter 2 outlined the kinds of transnational work that were necessary to allow or even attract particular forms of globalized environmentalism (i.e., international nature conservation organizations) to China, this chapter explores how the environmental winds were negotiated at the village level. Here we see how rural people and conservation staff used the art of engagement to respond to and remake environmental projects in ways that were far more nuanced and textured than theories of resistance or accommodation to globalization might have us believe.

XIAO LONG'S HISTORY OF ENGAGEMENT

Before the Mao era there was relatively little direct governmental presence in the uplands, and the closest center of power was in the present-day prefectural capital of Jinghong, previously the seat of the Buddhist, Dai-ruled Xishuangbanna Kingdom. The history of upland Banna remains to be written, but oral accounts suggest that, with the exception of taxes and regulations on trade in lucrative goods such as tea, salt, and opium, there was relatively little centralized control over upland villages.[15] Armies from northern China occasionally set up garrisons, but many soldiers succumbed to little-known diseases, and the area was long seen as a site of pestilence and sickness.[16]

Since the beginning of the Mao era, the uplands were caught up in waves of state campaigns: officials amalgamated villagers into communes, enjoined them to produce huge quantities of grain, and compelled them to build dams and irrigation canals. Throughout the 1960s officials relocated many small upland villages during a campaign to "bring the villages down from the mountains" and move them to bigger sites, especially those with more potential for paddy rice and closer to main roads.

Older residents described how they saddled up half a dozen oxen (*huang niu*, literally "yellow cows") and wandered the mountains, trying to find an empty valley that had good soil for rice and good *feng shui* for graves and would be approved by local leaders. Soon after they found a site, dark with thick trees, a platoon of soldiers came and sent them away, back to a temporary site. The soldiers came to support the Vietcong against American soldiers and left several years later. Some of the original settlers returned, only to evacuate their homes again for another group of soldiers who used the area as a base camp when China went to war against Vietnam in 1979. These two sets of soldiers dramatically transformed the landscape. They constructed rice paddies, terraced many hectares of tea plantations, raised cattle and pigs, and planted fruit trees, such as oranges, which were still struggling along in the 1990s. The soldiers used swidden agriculture to grow corn, used mainly for feeding animals, for it was considered a poor substitute for rice. With overhand hoes and no machinery, the soldiers carved out an eight-kilometer dirt road from Xiao Long to Xishuangbanna's main paved road, which had also been built by the army, in 1953. After the soldiers left Xiao Long, the commune leadership was concerned about the fate of its tea plantation, and it encouraged more people to settle. By the time all of the new residents were relocated, Xiao Long's tea bushes were badly neglected; they "had grown into trees and were taller than a house." The tea trees were chopped down in order to produce new shoots for harvest. This was during the Cultural Revolution (1966–76), which initiated a decade of dramatic change—often called "ten years of chaos"—in many parts of China, but less so here, where no Red Guards or "sent down youth" came.[17]

When middle-aged people talked of the past, they often mentioned the early 1980s as the key period of social transformation; this was the era throughout China when cadres came to dismantle the communes and village leaders distributed land to families. In Xiao Long, like many places nearby, the land was divided more or less evenly into three agricultural grades (mediocre, good, and highly desired land), based on considerations such as accessibility to reliable irrigation, soil quality, slope, and distance from the village. In almost all cases, families got one or more plots from each of the three categories; the amount was based on the number of people in their family, often an extended family with at least two children and paternal grandparents, as many places tend to be patrilocal (whereby the wife moves out of her own village into that of her husband's and lives with his parents). Unlike other communes in the province that had valuable collective resources like tractors or sewing machines to redistribute or sell by auction, there was little of much worth except some plows and oxen. There had been some attempts to raise pigs communally in Xiao Long, but this did not last long, as the pigs escaped too often.

Local residents were just getting used to the new changes when, in the mid-1980s, two groups of officials came, one after the other, bearing maps. One group showed the villagers the new boundaries for a nature reserve. The next group brought a map showing how village forests previously designated as subsistence lands for farming and firewood would be added to the reserve. In 1987, the year before WWF arrived, staff from the Nature Reserve Bureau visited Xiao Long bringing with them a new concept: they told residents not to hunt "protected species."

Villagers already knew the central government valued the lives of one species, the wild elephants, but this was the first time that reserve staff had said

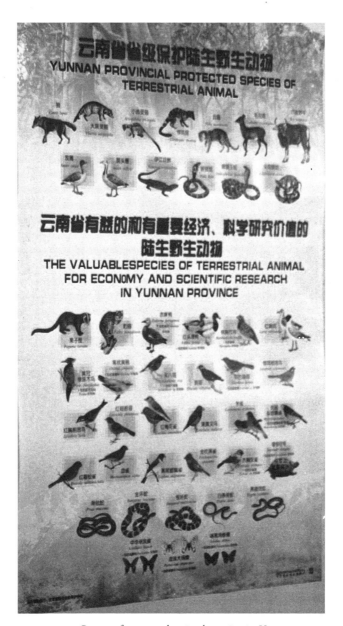

FIGURE 10. Poster of protected animal species in Yunnan, 1999. Xishuangbanna, Yunnan. Photograph by Michael Hathaway.

that several other large mammals, like gaur (*ye niu*, a kind of wild cattle; *Bos gaurus*), should not be killed.[18] Gaur had been a favored target for groups of soldiers and others. Thus even before WWF arrived in 1988, villagers had already begun to feel the impact of the environmental winds as state representatives created new boundaries between farm and forest and began to criminalize practices like hunting, which they had virtually ignored, and swidden, which they had encouraged.

In sum, it is important to understand that although many accounts of development encounters largely ignore the previous history of domestic and international engagements in particular places, in part to frame their account as a novel incursion, the people of Xiao Long had long been involved in a wide range of efforts to transform their lives and the larger society. Through these experiences they had learned how to negotiate a number of powerful winds—how to seek shelter if they could, but also how to work with and manipulate the winds, which is part of the art of engagement.

ENGAGING WITH WWF

When WWF arrived in 1988 their primary goal was to address tropical deforestation. To that end, their development plans were aimed at separating villagers from the forest and encouraging them to increase market participation, an approach that was de rigueur for conservation projects around the world in the late 1980s (Peet and Watts 1996). WWF staff encouraged villagers to construct fish ponds, plant tree lots for fuel wood, cultivate fruit trees, and, most important, implement agroforestry—the deliberate interplanting of annual plants and trees. WWF also created projects incorporating resident wild elephants, China's last herds. As I explore in chapter 4, WWF staff built China's first ecotourism center for viewing wild elephants and brought in electric fencing to help defend villagers against these elephants, their worst agricultural pest.

Although WWF is one of the most powerful conservation organizations in the world, working in dozens of countries and with an annual budget of over U.S.$100 million, its Xishuangbanna team was quite modest: it consisted of Jack Bentley and his Chinese assistant, Yang Bilun. Bentley, an energetic Euro-American in his late fifties, had international development experience in the Philippines. Yang, in his early forties, was a trained rubber specialist with a strong background in biology; he had acquired excellent

English skills and felt a strong commitment to rural welfare. He too was an agroforestry missionary. Despite the small size of the WWF team, local residents took their project very seriously.

When Bentley and Yang arrived in Xiao Long they were accompanied by a handful of staff from the Xishuangbanna Nature Reserve Bureau, most of whom were reluctant to leave their offices to begin a village-based project. For the next seven years, Bentley and Yang, sometimes accompanied by reserve staff and sometimes not, would arrive for periodic visits of a week or more. Although Bentley had powerful connections to outside forces such as WWF and organizations in America, most residents quickly realized that he was in a precarious position vis-à-vis the Nature Reserve Bureau and Chinese state officials. He had no authority to enforce any mandates, and so he mainly relied on persuasion, trying to convince villagers that his plans were viable. He found that residents enthusiastically embraced some aspects of his plan. When he encountered reluctance to plans in which he had particular interest, he tried to rally support from nature reserve staff, who, as state workers, had some capacity to pressure or compel villagers. To some extent, Bentley had potential leverage with the nature reserve staff, and he was able to capitalize on the fact that WWF had played a significant role in the Nature Reserve Bureau's rapid rise to more power and influence, at least within their "kingdom," yet WWF had no power to create or enforce any laws. Bentley himself, however, had little to do with this rise in influence, and Bureau leaders felt more obliged to the head of WWF than to Bentley himself, who was known as a short-term contract worker. Even though some reserve staff appreciated WWF's international status, they were also wary of alienating the villagers and thus not always willing to cooperate with Bentley.

By the time Bentley arrived, the Nature Reserve Bureau staff were in the middle of a significant transformation. In the early 1980s they were largely low-status and poorly paid workers with little political power in a chronically underfunded and understaffed department, who rarely traveled from their small offices. Bureau managers leveraged Prince Philip's visit and WWF's project to make an argument to top officials in Beijing that their bureau should be promoted from a prefectural to a national rank. They were successful, which meant that the Bureau leader gained a high position in the regional government hierarchy, almost equivalent to the mayor, and thus achieved an influential role in shaping regional politics. The leader also convinced WWF to donate several vehicles to the Bureau at a time when cars and jeeps were very difficult to obtain in China. These vehicles were a critical status symbol and enabled the staff to do much more than visit reserves and monitor villages. Cars facilitated their ability to cultivate social networks with local government officials. Even though the highest levels of WWF staff questioned the necessity of such a gift, lower-level WWF negotiators tried to obtain cars to give the Bureau leaders, Although there were tensions between WWF and the Bureau, the Bureau leaders greatly appreciated these material goods and symbolic support.

Beyond WWF's material gifts, the Bureau gained much power as a result of new environmental protection laws, which they were entrusted to enforce. Each new law gave them potentially more responsibilities as well as more discretionary potential. In less than a decade during the 1980s and 1990s, the power of the Bureau increased dramatically, and there were more intersections between the lives of Bureau staff and Xiao Long residents as the Bureau gained vehicles and started more nationally and internationally funded projects. Bureau staff became the most frequently seen government representatives in Xiao Long and other villages within the Bureau's kingdom, consisting of thousands of villagers over a massive land area. They promoted and mediated development projects, made and enforced laws, and provided payments to farmers for crop damage from pests like elephants. The Bureau also became a kind of landlord and taxed residents for planting sha ren, which they had begun to cultivate under the canopy of the nature reserve's forest, with the blessing of regional officials.

Whereas urban citizens often experience the state as an abstract and anonymous bureaucracy, Xiao Long residents were like rural citizens elsewhere, who often experience the state as embodied in a few well-known individuals, such as tax collectors, police officers, and agricultural extension agents (Corbridge et al. 2004; Gupta 1995). Just as the state was represented by specific groups and individuals, Bentley was seen to represent WWF and the desires of "the Americans."[19] As I discussed earlier, most villagers tried to cultivate new and better relationships with state officials, Bureau staff, entrepreneurs, and itinerant development workers like Bentley. In contrast, Bentley saw himself as on the "side of the villagers" and against officials, whom he tried to avoid as much as possible.[20]

Bentley and Yang tapped into a rapidly expanding network of agroforestry researchers. This network was based in Kenya, at the International Center for Research on Agroforestry (ICRAF), and had initially begun with agricultural extension and soil conservation agents in the developing world, though it later included conservationists and missionaries. Bentley and Yang emphasized a method pioneered by the ICRAF called "alley cropping," which was usually practiced on terraced hillsides, with nitrogen-fixing trees at regular intervals in rows or "alleys" adjacent to grain crops. The trees' leafy branches are periodically clipped and added to the surrounding soil. These trimmings and the trees' roots provide a supply of nitrogen, the world's most common commercial fertilizer, to nearby crops.

Before he came to China, Bentley learned about alley cropping while working for the Sloping Agricultural Land Technology (SALT) project, run by evangelical Christians in the Philippines' uplands.[21] The WWF's *Agroforestry Handbook* portrays their vision of an ideal landscape: terraced hill slopes planted on contour lines and strips of annual grain crops alternating with perennial nitrogen-fixing trees. The mountaintops and ridges (thought to be most at risk for severe erosion) were covered with trees, and there was a clear separation between cultivated and wild lands.

Bentley found many challenges in promoting agroforestry. He did not, and could not, simply impose his vision on Xiao Long residents. Yet he saw agroforestry as a clear path toward bettering their lives and promoted it with zeal. He also saw himself as an advocate for villagers, whom he imagined as surviving under the dictatorial mandates of an uncaring state. His antistate views were not far from those of his neighbors near his farm in the western United States, and they were increased by a particular understanding of China as authoritative and officials as elitist. He criticized what he saw as a trend toward increasingly strict state-led conservation efforts, which were part of the environmental winds. He worked to shift the direction of the environmental winds toward what he saw as a more peasant-friendly, intermediate approach, such as providing technical training, seedlings, and support for agroforestry.

Bentley was quite disturbed that the Xishuangbanna Nature Reserve Bureau had relocated over eight hundred people between 1987 and 1990 (Yu 1993). This was despite the fact that WWF (1996: 9) as an organization had encouraged the relocation, viewing it as "indicative of the seriousness of the attention paid to conservation by the local government."[22] Bentley wanted to visit the new village sites but was denied permission. He believed there was no way to protest the government's decisions, which were, he said, "always final in China." Some of the old village sites were burned down to ensure that no one returned.

With new restrictions, Bentley claimed, villagers' "old ways of slash and burn," or swidden agriculture, were no longer viable. He believed it best to

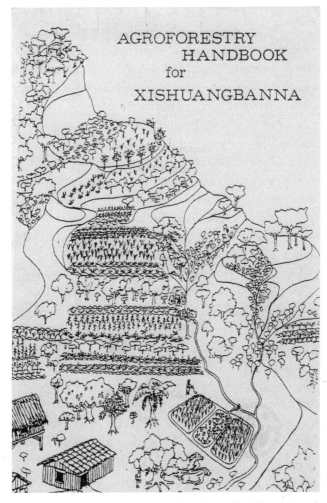

FIGURE 11. Cover of WWF's *Agroforestry Handbook* (Bentley 1990) showing crops on contoured terraces.

help villagers use their lands intensively by planting tropical fruit trees, grafting productive varieties, and practicing "proper tree management." He expressed interest in tropical forests and a concern for soil erosion, but these were clearly not his only concerns.

He used the terms *ethnic minorities*, *farmers*, and *hunters* to refer to local inhabitants and described himself similarly, as a farmer, a hunter, and an ethnic minority. He criticized the condescending attitudes held by Han-identified Nature Reserve Bureau staff and was particularly bothered by what he saw as the government's persecution of ethnic minorities, including the people of Xiao Long. His views stood in marked contrast to WWF's earlier project reports written by his supervisor John MacKinnon, who described upland minority groups as particularly ignorant and environmentally destructive. Bentley did not see them as "indigenous people" with internationally recognized rights or try to articulate their cause with indigenous advocacy groups, but he said that his own status as part Cherokee Indian helped him understand their plight as fellow ethnic minorities.

Bentley believed that his project would be successful because of his close connections with local people, based on his "farmer's sensibility."[23] In everyday conversations and in his writings, Bentley referred to those he worked with as "villagers," not "peasants," the latter a common term in China, by the 1980s rife with patronizing associations (Cohen 1993). Critical of Chinese officials, he reinforced his allegiance with villagers by repeatedly and overtly distinguishing himself from officials. In his final report for WWF, he explained that he carried a long knife, which he sharpened during village meetings: "They nod their approval of the knife and my sharpening as they return the knife to me. They understand I know how to use a working knife. As we walk around their farms, I never pass a protruding branch that might snag a load some weary farmer is carrying. I reach up with the knife and cut it off. The farmer sees this and understands that I know about some of their basic burdens in farm life" (Bentley 1993: 31). He then juxtaposed his behavior with a "member of the official party" who did not attempt to cut or break off the branch, who either does not know or does not care about the "basic burdens of farm life" (31).

Bentley's village experiences were deeply constrained, however, by language and time. He did not speak Chinese, and even though he always worked with a translator, many aspects of his time in the village were never translated or explained. He followed state regulations that prohibited foreigners from staying overnight in rural villages. Therefore he and Yang slept at an approved hotel in a town and commuted over an hour to Xiao Long, arriving after breakfast and departing before dinner, usually in a Nature Reserve Bureau jeep.

Although Bentley expressed his sympathies for rural residents, his position as WWF's agroforester required that he also attempt to remake their social landscape. In this role, he justified WWF's goals of social transformation as outlined in the project design and his own plans for agroforestry. In order to argue the need for agroforestry, he criticized their existing agricultural methods and other livelihood practices. He created a booklet with photographs of local agricultural practices he said were destructive and provided prescriptions for their solution. As much as Bentley's position with WWF enabled his potential to assist the villagers, his job also placed him within larger constraints of WWF policy.

The art of engagement is always conducted within not fully comprehensible fields of constrained choice, influenced by complex social relations one does not always know about or understand or have the capacity to change. Efforts to connect with particular groups or individuals may fail, or if they ostensibly succeed, these relationships can produce unintended consequences or even jealousy where none existed before. Bentley felt hemmed in by his role as a WWF consultant and caught between Chinese intergovernmental conflicts. One of many conflicts occurred after WWF had their project approved by the central-level Ministry of Forestry. WWF was supposed to collaborate with the Ministry's provincial-level representative, the Yunnan Forestry Bureau, but instead forged stronger ties to a forestry college located lower down the bureaucratic hierarchy. This produced institutional jealousy and upset officials at the Ministry of Forestry, which meant additional difficulties for Bentley. He mediated institutional resistance to his project and struggled to obtain a visa during his periodic visits to China; the Ministry, he claimed, intentionally delayed the process. At other times, resistance to his project manifested in outright conflict with officials.

These conflicts emerged in particular ways and were notably mediated by a lingering Cold War atmosphere, still palpable during my fieldwork.[24] As mentioned earlier, a number of officials knew that Bentley was a Korean War veteran, as were some of them. In another case, some officials (likely from the disgruntled Yunnan Forestry Bureau) spread rumors that Bentley was a spy.[25] Midway through the project, Bentley secretly arranged to pay some Chinese villagers, who frequently crossed over the Sino-Vietnamese border, to inquire into stories that some American prisoners of war remained in Vietnam, doing forced labor in a mine. The provincial government learned of his inquiries. As one of my colleagues explained, Beijing's relations with Vietnam were still sensitive after the two countries' military clash in 1979; officials were disturbed at the idea that Yunnan was harboring an American with military experience who paid Chinese to secretly investigate conditions in Vietnam. As far as I can tell, Bentley was never aware of such accusations, but nonetheless it had ramifications. In China resident foreigners are assigned to a state work unit that is responsible for their actions. As such, the college came under greater scrutiny and was rebuked for allowing him to interfere with and potentially harm Sino-Vietnamese relations. As much as the college reached out to outsiders, including international organizations, and as much as the art of engagement is necessary for potential gains, there were also many unknown constraints and risks, such as being responsible for Bentley's actions.[26]

THE DEMONSTRATION PLOT AND THE FENCES: BAMBOO AND BARBED

With plans in hand, Jack Bentley and Yang Bilun plunged into the difficult task of getting the project started. Although the term *agroforestry* was given the literal translation *lingnongye,* villagers it called *da nongye,* or "great agriculture," using the term *great* to designate its seriousness and also reminding people of other uses of the term, such as the Great Cultural Revolution (*wenhua da geming*). Bentley tried to persuade villagers to engage in a number of activities. Many plans, such as cultivating a large plot of trees for fuel wood and growing nitrogen-fixing trees in rice paddy fields, were rejected as impractical by Xiao Long residents. Their ability to reject Bentley, however, was not absolute. For WWF, a key and nonnegotiable component of the project was the establishment of an agroforestry demonstration plot. It would require a series of terraced plots with nitrogen-fixing trees interplanted with corn. For local residents, the most daunting aspect of agroforestry was that it would require vast amounts of their labor to terrace the hills. This represented a physical and temporal burden to villagers unconvinced that the benefits of agroforestry would outweigh their efforts; they typically saw trees as competing against crops, so planting them together didn't seem wise. They were already quite busy expanding their cash crops (e.g., sha ren) in the forest, leaving them little extra energy to spend terracing fields for corn.

Moreover their view of this project was inextricably linked to memories of past campaigns. Since the end of communes in the early 1980s, residents' work on group projects (such as road repair) earned them money or credit. Even during the commune period, labor earned work points, which were exchanged for grain and money. The WWF plan, on the other hand, represented the first time they were expected to work without any compensation, beyond creating a potentially larger corn harvest.

The Nature Reserve Bureau committed to working with WWF to ensure that the demonstration plot was carried out. Bentley wanted the plot to be of a certain size, location, and slope. He wanted the site to measure one hectare so that productivity figures could be easily derived from the agroforestry experiment.[27] This would enable him to later compare levels of productivity inside and outside of the plot. There were a number of criteria for the demonstration site:

The first plantings should be:

- where efforts are most likely to succeed
- where the young trees and/or shrubs can receive adequate weeding and watering
- located where others can easily observe and participate. (Bentley 1992: 2)

Thus the site should be close to the village center, where it would be highly visible and demonstrate the effectiveness of agroforestry. As Yang stated in 2000, they also wanted to establish the site on sloping land in order to imitate slash and burn, the form of agriculture it was designed to replace. Yet when it came to choosing the actual site, Bentley and Yang were worried that the steep slopes of many former slash-and-burn sites might be too susceptible to erosion and water runoff, even though these were precisely the problems that agroforestry was supposed to solve. Above all, Bentley and Yang wanted the agroforestry plot to be a success and argued that a flatter slope would cause fewer problems. This desire for success drove them to find a site that would not definitively demonstrate to villagers how agroforestry could replace steep-slope slash and burn, which was, after all, the major aim of the project.

Instead of choosing an old swidden site or a plot on a steep slope, the Reserve Bureau staff chose some of the last remaining land on a gentle slope that was already terraced. Tea fields planted by soldiers during the 1960s and 1970s surrounded the demonstration site. In the chosen plot, many tea bushes had died back, providing space, and the soil was still relatively fertile. Xiao Long residents had already planted corn in some areas. Although the plot did not demonstrate what it was intended to, it did meet Bentley's logistical requirements, the Bureau staff's contractual obligations to WWF, and Xiao Long residents' desire to avoid the heavy labor of constructing terraces from scratch; it was therefore a relatively successful form of engagement, neither a perfect solution nor a burden shouldered by one group, but somewhat satisfactory to all parties. Yet even creating this simple plot required much work and effort. Initially the plot, which contained the land of ten families, was not particularly valuable, as it was not irrigated. As the project progressed, however, the plot increased in value due to its protective fencing.[28] Once the demonstration plot was selected, the Bureau staff commanded the villagers (*rang nongmin*) to construct a bamboo fence. Under the guidance of Bentley and Yang, residents planted the seedlings of some of global agroforestry's most cherished nitrogen-fixing species, *Flemingia macrophylla* and *Gliricidia sepium*, as well as other quick-growing tender perennials such as Pigeon Pea (*Caganus cajan*) and Sesbania (*Sesbania bispinosa* and *S. sesban*).

Obtaining these plants was quite difficult, as these species were ones favored by the main agroforestry research center in Kenya and seeds could not be found in China, let alone seedlings. Bentley finally tracked down a seed supplier in India, where they were purchased and shipped at great expense. The first batch lost its viability after spending a long time in quarantine at the customs station in Shanghai. Before ordering the next batch, Yang pushed his guanxi network to the limit to find someone with connections to a Shanghai customs officer who could keep an eye out for the next package and make sure the seeds were taken care of and quickly sent on. Through this newly created connection, Yang ensured that the customs time was reduced. After several more months, the second batch arrived, and the seeds, to everyone's great relief, were still viable. Bentley and Yang organized a nursery to raise the seedlings, which seemed to grow quickly and herald a positive turn in the project's future.

A few months after the seedlings were planted, water buffalo broke through the bamboo fence, eating many seedlings to the ground. After Yang discovered this, he drove back to town and alerted the nature reserve staff. Within the week, the staff brought the first rolls of barbed wire to Xiao Long. The next day, they told each family in Xiao Long to contribute a few stout poles for fence posts. Men went to the forest to cut decay-resistant species and brought back the posts. Women dug the holes, tamped in the posts, and strung the wire. The fence required dozens of trees, causing one resident to comment that, in the end, the project consumed (*chi*) more trees than it raised, even though, as another resident put it, "Bentley really seemed to love trees."

I learned that this was one of the main incidents that led villagers to reflect critically on their water buffalo and, over the next decade, to replace them with hand-held, two-wheeled tractors. Bentley himself did not see buffalo as a key problem, but he was frustrated that villagers seemed to take either a casual or an environmentally destructive approach (by digging trenches) to protecting their crops. After his project was over, villagers debated the reasons WWF abandoned them, and the water buffalo were mentioned. Just as villagers developed plans for future roads and made a list of potential village projects in case the opportunity arose, they also got rid of buffalo, an important part of their lives. Many of the older men still mourned their loss, as they had grown up and lived with buffalo their entire lives, spending their youth accompanying them while they grazed and riding on their backs. They knew so much about them, including the buffalo's "one hundred sicknesses" (*yi bai bing*) and their cures, whereas they knew nothing of the hand tractors and could not repair them when they broke. Yet the younger generations won the decision, inspired by their desire that in the future they could be connected to WWF again, or some other conservation group.[29] Hoping for these connections was not their only consideration, as the buffalo had also created social friction when they ate others' crops, and they predicted that this act would also be regarded favorably by local officials, who saw the tractors as emblematic of their village's financial success and eagerness to embrace symbols of modernity. Being seen as a progressive village had many advantages in China, for these were the places that were more frequently given new development projects or became part of study tours for visiting officials and could thereby gain more connections with local cadres. In any case, getting rid of buffaloes showed that Xiao Long residents were willing to make major changes in an effort to foster more forceful winds, and they did so before their neighboring villages.

RETHINKING CHOICE AND PARTICIPATION

Bentley's project was a continual exercise in the art of diplomacy, not the power of imposition. Like all forms of engagement, his project was to some degree mutually constructed, neither fully accommodated nor resisted. He gave up on a number of plans after he faced overt criticism or simply a lack of enthusiasm. For instance, he was inspired to learn that in the lowlands of Banna there were ethnic Dai villagers who grew their own "firewood forest" with Cassod trees (*Cassia siamea*), a fact he learned by reading English-language articles by Yunnan ethnobotanist Pei Shengji (described in chapter 3). This practice fit well with conservationists' imperative to keep villagers separate from the wild forest by reducing their "unhealthy dependence" on it. For Bentley, adopting the Dai practice of raising trees for fuel wood seemed like a form of "local knowledge," a concept that was just coming into vogue among development groups. Xiao Long residents stated, however, that it was not *their* knowledge or practice. They had experimented with a few Cassod trees, but they said Xiao Long was too cold in the winter for the trees to prosper, and there was really no need to grow firewood when there were millions of trees all around them. Bentley quickly abandoned the idea of growing fuel wood but maintained his key goal of creating an agroforestry project. To do this, he relied on the assistance of the Bureau staff, who had a greater ability to command villagers. Villagers rarely enlisted because of threats of discipline, but neither did they choose in a social context free of consequence. Li Wen, who owned land in the agroforestry plot, clarified this point for me.

I asked Li if she had a choice whether or not to commit her land within the plot to the agroforestry project. She said that she could choose not to be in the project, but that because the agroforestry plot boundaries were fixed, she would have had to trade her land (within the designated plot) with someone else's land (outside of the plot). "But, they said it was best if we didn't trade. The nature reserve people came and brought in a contract [*hetong dingxiale*] for me to sign." She told me that villagers often traded land but that no one dared trade land in the zone designated as WWF's demonstration plot, in part for fear it might anger the reserve staff.

Li was more concerned about the future consequences of acting against the wishes of reserve staff than she was that her land might be squandered. Indeed the staff's presence had increased over the years, and they could play an important role in the lives of villagers bordering the reserve, then and for the foreseeable future. WWF, however, was seen as a transitory agent. Villagers recognized that, long after the project was over, reserve staff would continue to be involved in important decisions that exercised some level of personal discretion. Staff could decide whether and how much to compensate them when a state-protected wild animal (such as an elephant or a wild ox) ate their crops, and whether to jail them and subject them to stiff fines if they were caught hunting or farming on the reserve. Given this, Li neither resisted not embraced the project goals but carefully negotiated her long-range possibilities within a given set of options and constraints. I heard a similar perspective when her father shrugged and said that when he had worked on several dam projects during the Great Leap Forward, he felt a desire to do the work but also lots of pressure to volunteer.

As much as the residents of Xiao Long were leery of reserve staff and reluctant about many of Bentley's plans, many expressed hope about what the project might bring. Like the state's development projects, the WWF's had the potential to bring hardship or assistance. The project seemed to have much to offer, such as travel opportunities, material goods, and other unknown opportunities. Early in the project, WWF bused a number of residents to a series of workshops, provided them with free meals, lectured them on agroforestry, and paid for their visit to the Xishuangbanna Tropical Botanical Garden. This kind of treatment was without precedent, and some said they "felt like tourists" for the first (and last) time, touring in a kind of luxury they rightly felt would not continue for very long. They knew that, like being a good village in the eyes of local cadres, there were advantages to being in Bentley's good graces and that his project could bring all kinds of novel possibilities. Thus many tried to include him in their own social networks, as much as they found his perspectives and plans somewhat confusing or daunting.

But in other ways, the WWF project followed residents' expectations about state-driven development projects. As rural Chinese citizens, Xiao Long residents were familiar with unpredictable and sometimes heavy government demands on their labor and livelihoods. This had especially been the case case during the Mao era. These experiences gave many of the older residents a cautious attitude toward new campaigns, and they were less likely to quickly embrace new plans for changing their livelihood. WWF's actions of claiming property outright (the agroforestry demonstration plot), purchasing it under duress (for fish ponds),[30] and demanding villagers' labor (such as for building the fence around the plot) were familiar. Although WWF later labeled the project as "participatory," it relied more on residents' participation as free laborers than as equal collaborators.

In all, the local residents' engagement with the project was inflected by how much the Bureau staff pressed certain activities. Although many of Bentley's requests were easily dismissed by villagers, those that the state promoted, such as the demonstration plot, were difficult if not impossible to refuse. As we saw with Li Wen, however, compliance was not merely exerted through intimidation; it was also motivated by villagers' interest in cultivating positive long-term relations, especially with Bureau staff (who they knew would remain important state representatives in their life), and also with Bentley, who seemed to arrive out of nowhere, to which it was generally expected he would return. Thus while their response to certain elements of his plan could be seen as resistance (e.g., families with land in the agroforestry plot, despite their "reeducation" in understanding how to carry out agroforestry, continued to work on distant swidden fields rather than abandon them), this was not their only or main response, but one of many. The important point, I think, is to recognize that while the art of engagement can certainly include actions that may qualify as resistance, it can just as often include efforts to foster social relations, lessen the impact of potentially harmful plans, and discover and create new opportunities. Whereas theories of resistance assume a uniform response to dominance, I found that some families engaged much more than others, positioning themselves differently in regard to Bentley and Bureau staff. Some were willing to take more risks with their tried and true agricultural practices and experiment with the various ideas Bentley was proposing. Their engagement was not simply a matter of working extra hours to plant and care for trees; it also came through in the other ways they tried to gain his favor, such as going far out of their way to prepare meals for him. Many described to me how some families had worked hard to cater to Bentley's strange culinary habits, which included his astonishing refusal to eat rice—something that defined him as a curiosity more than anything else.[31] Whereas many scholars have shown how opportunities provided by development projects are often captured by the rural elite (Chambers and Leach 1989), in this case Bentley did not create close connections with the wealthiest members of Xiao Long. Especially for those who did not have many financial resources, using the arts of engagement to cultivate relationships was one of their only ways to get ahead.

It should be clear that Bentley had difficulty implementing aspects of his project and often had to deal with villagers' reticence. But in other ways, villagers reached out to him, grabbed on to some components of his project enthusiastically, and tried to use them in novel ways. Bentley, we should remember, was also practicing the art of engagement; to do so, he sought out particular allies in Xiao Long. He hoped to find some diligent "model farmers" who were also respected citizens, who could highlight the benefits of his methods. And some villagers did in fact "adopt" him by showing interest in what he said, or cooking for him, or just chatting (*chui niu*). A few residents were eager to gain material benefits, such as seedlings, and free training on growing fruit trees.

GROWING FRUIT TREES FOR WWF

In some ways WWF presented an interpretive puzzle to villagers, but in other ways it was similar to state-based efforts at development. During my conversations in 2001 with Xiao Long residents, many still wondered about Bentley's interest in growing fruit. Were the residents meant to grow the fruit for themselves or for sale to others? What did fruit trees have to do with conservation, which seemed mainly about protecting wild animals? Prior to WWF's arrival, Nature Reserve Bureau staff mainly lectured villagers about not shooting large mammals. Villagers saw WWF's panda icon (which seemed strange to them, as they knew it was a foreign organization, yet it used an animal exclusive to China), so it also seemed WWF prioritized large mammals. If this was their main goal, what did this have to do with growing fruit?

Some people ventured that WWF shared the government's desire for rural development. Lu Wen, a young man with an entrepreneurial streak, noted as much: "Like the government, Bentley wants us to plant fruit trees. He wants us to build a production base." Indeed WWF's fruit tree recommendations were similar to government campaigns and promoted chemical fertilizers and pesticides for increased yields. I read in the archives that WWF gave a backpack tank sprayer for pesticides to each family involved in the project. Thus, like the state, WWF promoted increased pesticides in order to control insects and disease and to produce larger and more appealing fruits, perhaps for the market. Yet, as Lu Wen opined, this venture could not work if only a few people participated in growing fruit trees. Only if their village made a "production base" with a sufficiently large harvest of fruit could they attract dealers to make the trip down their dead-end dirt road, an issue that Bentley never raised. They had already faced challenges in attracting tea dealers willing to pay a good price; their processed tea was stored in a brick room, where it could be safely held for months. Tea was easy to ship, but fruit was finicky. Villagers saw their dirt road as a major impediment and worried that their upland location, with occasional freezing winter winds, was too risky for tender tropical trees.

Purportedly Fu Ganxing was the only one in Xiao Long who tried to market tropical fruit grown with project support. Fu was in his fifties, well-dressed, and very energetic. But, as I soon found out, he had what was called a *maodun* (contradiction) with the others in Xiao Long. He followed the project's directions meticulously, pruning and removing dead branches with small hand pruners. In 2000 I examined the old agroforestry plot with Yang Bilun, and we talked with Fu, who was there trimming his *long yan* (dragon eye) fruit trees. Fu asserted that he was "the only one in Xiao Long who remained committed to the project."

His claim of exceptional commitment paralleled his high degree of social isolation. Fu was relatively wealthy from his sale of tea. He leased a large plot of tea bushes from leaders of an army platoon that had been based in Xiao Long and maintained some land claims. He built a small tea factory in a twenty-by-thirty-foot building with split bamboo walls and a thatched roof and hired outside laborers to pick and process the tea. Many in Xiao Long envied his wealth and criticized his separatist ways. According to some, Fu tried to court Bentley and Yang. In 2000 Yang told me that he and Bentley first considered asking Fu to become one of their model farmers but then realized that Fu's peers did not respect him, so they refused his advances. Had Fu been on good terms with others in the village, it is likely he would have become a model farmer, but his social standing worked against his attempt to engage.

Some residents selectively experimented with Bentley's tree recommendations. For example, a young man named Li Bo explained that Bentley's demands for pruning the trees were very particular: "Bentley wanted the cuts to be neat, close to the trunk." Before Bentley's arrival, almost all of the work on fruit trees was done with machetes, used to lop back branches, but Bentley wanted them to use the hand pruners he brought. Li Bo and some of his friends compared the health of trees trimmed by Bentley with pruners and by their own machetes; as there seemed to be no difference, they didn't follow Bentley's advice. Li's statement belied the frequent assertions of domestic and international experts that they provided tested scientific knowledge and that locals merely believed in untested folk understandings. "We experimented. Bentley told us what he thought was true, but it was probably just true for the United States. Here it didn't matter." Li subjected Bentley's statements to empirical tests and even afterward did not entirely dismiss Bentley's ideas but rather explained his own idea that some of Bentley's knowledge, like other forms of knowledge, including his own, might be specific to place. Li neither flatly rejected all of Bentley's knowledge claims nor embraced them as true, just as he evaluated Bentley's plans with a skeptical but largely open mind. This attitude was often frustrating for both Bentley and Chinese agricultural extension agents, who wanted villagers to quickly adopt their plans without questioning them. Yet this stance likely served villagers well, providing a framework to guide how they selectively engaged with outside ideas while working on their own plans.

Li's critical yet open-minded approach did not mean that he often challenged Bentley, and certainly Bentley became close to his family. One day Li showed me, with a bit of pride, that he still owned a pair of pruners that Bentley had given his father, and there was a sense that members of his family had been willing to take some risks with the hope of Bentley's approval. Yet they also had different aims. Li made it clear to me, as he had not done with Bentley, that most people in Xiao Long weren't particularly interested in growing fruit trees because they did not believe that fruit would bring them sufficient returns, as prices were low, and their long, muddy road would mean great difficulties in bringing the fruit to market. Instead they were interested in loans to build a new house, buy a walking tractor, raise fish, or open a small store.

Yet all of these plans required them to be connected to places beyond the village, and there was a deep sense of worry about being "left behind," a concern exemplified by a pulsing boom from far away that lasted for months, explosions from the construction of a new highway being cut through hills. This road would not make Banna much more convenient to Kunming because it redirected traffic to another valley, far from the old highway that connected them to their own spur road. The village had worked out a series of rough sketches for constructing a road that would connect Xiao Long to the new highway. After the highway was built, how could they attract important people, like the tea boss, who came to buy their processed green tea? In showing me their plans, they said that a group like WWF might pay for their new road so that tourists could use it to see "real wild elephants" (unlike the trained elephants in a nearby tourist center).

This discussion got them talking about all of the money that was rumored to have gone into WWF's project. They had unsuccessfully lobbied Bentley to assist with access to these loans. These forms, however, diverged from WWF's interest in promoting conservation-oriented horticulture. State and international agencies often imply that "target villages" like Xiao Long are supposed to express gratitude in receiving development projects—as if they were a gift. Yet residents still actively debated how WWF and the Nature Reserve Bureau used the project money. Some said that much was squandered and that they themselves would have made better use of the funds. Villagers did not see this project as a gift but questioned its rationale, debated its cost effectiveness, and evaluated it in relation to other alternatives.

Along with Fu Ganxing, one other man was enthusiastically drawn to the project's support for fruit tree cultivation. He took selective advantage of WWF's offerings. His nickname was "Guangdong," after his home province, which he often remarked was much warmer than this mountain village. Often cold himself during Xiao Long's winters and bundled in sweaters and a knit cap while his neighbors wore shorts and flip-flops, Guangdong was worried about the periodic winter frosts. Unlike others, he did not buy WWF's subsidized tropical fruit tree seedlings. Instead he bought cold-tolerant plum seedlings using his own guanxi network, mainly consisting of people from his home province. Using this same network, he found a fruit broker to harvest and sell the fruit. Otherwise, Guangdong said, to pick the fruit, bring it to town on the bumpy road, and sell it himself at the market would be a risky and difficult venture. Even if the fruit was not bruised by the trip, it would spoil quickly.

Guangdong's interest and relative success with the fruit trees highlights several points. First, those most attracted to these development projects were not the average citizens of Xiao Long but socially or economically atypical men. Marrying into Xiao Long, Guangdong did not inherit land. Like Fu Ganxing, who owned the private tea factory, Guangdong was the only other permanent Xiao Long man who did not cultivate rice. This freed them both up to focus their energies elsewhere, but it also meant that they had to make more money to buy their daily rice. Although Fu was able to invest in tea, Guangdong had scrambled for years to make a living.

Second, Guangdong's relative success came neither from closely following WWF's protocol nor from resisting it, but from using some of it and disregarding much of it. His use of the project entailed the careful sorting out of those aspects he thought practical and locally suitable and those he considered foolhardy; it was, in other words, a selective form of engagement.

Third, his use of the project relied on what he had already established: a rich set of social connections (guanxi) with entrepreneurs from his home region.[32] His investments in the fruit trees tapped these connections, as well as extended and deepened them, as he had more favors to ask as well as more to offer. Others in Xiao Long, who often knew far more villagers in the region but relatively few entrepreneurs, could not really engage in the fruit tree project the way that Guangdong could. Few were willing to risk the cold climate in growing tropical fruit trees and undertake the difficult collective work of creating markets just to please Bentley. Most did not feel that all the labor involved, even though the fruit trees were nearly free, was worth it.

Bentley, for his part, did not seem to recognize that ensuring viable markets was a key to fostering a local desire to grow these fruits. Guangdong took advantage of the project's classes on grafting and other technical subjects, but more important, he used his place-based networks to buy his own seedlings and to arrange for a wholesale buyer.

Like Guangdong, other Xiao Long residents realized that outside organizations such as WWF could bring supplies and training but that such organizations seldom altered the residents' conditions or livelihood in fundamental ways. For instance, even if Bentley provided them with excellent varieties of trees to grow fruit or cultivate firewood, they would still need to contend with the larger social and political order. In the former case, Guangdong's experience showed many of the larger issues that impinged on being able to sell fruit. In the latter case, they questioned why they should grow firewood in their old swidden fields when they were already surrounded by millions of large trees: firewood was already abundant. If they decided to grow and sell firewood, they would face a maze of government restrictions and high taxes on these sales. Regionally there already was a thriving firewood business, mostly for restaurants in urban areas, but this was controlled by wealthy businessmen who were able to evade such regulations and taxes. "Selling firewood is not for peasants," said Li Bo. In sum, Bentley could not offer them assistance in navigating that larger world. To substantially improve their condition, residents felt that above all they needed connections, especially to powerful Chinese outsiders, such as high-level government officials or dealers in tea, fruit, and sha ren. Some families became quite close with Bentley and benefited in myriad ways, but ultimately many people in Xiao Long discovered that long-lasting power to change their lives was more possible through Chinese, not foreign, social connections.

SUMMARY

WWF's project in Xishuangbanna was one manifestation of the environmental winds: in an effort to save tropical rain forests, conservationists attempted to reconfigure the relationship between people and the nonhuman world by keeping villagers on the farm and out of the forest, thereby reshaping social and physical landscapes. Although I originally thought that I would witness rural resistance against the state and international forces that were promoting this project, my study showed me that the ways in which people responded to the project were far more complex. Contrary to many accounts of subaltern resistance, these villagers were not antidevelopment, nor did they wish for autonomy from the state.[33] Nor were their reactions to an external agency largely passive and predictable or an attempt to fight for the status quo. Indeed all those involved in the project—villagers, WWF, and the state—were reaching out, making new connections, finding grounds of collaboration. In other words, they were practicing the art of engagement, which deeply influenced how the project unfolded.

Even before WWF had arrived, villagers were aware of the environmental winds blowing through their village, and they were actively reaching out to them as they dodged new threats and searched for new opportunities, all the while experimenting with a variety of new projects. Some of these projects were promoted by the government, some were suggested by entrepreneurs (*laoban*), and some were thought up by the villagers themselves: raising fish, growing passion fruit, planting new varieties of tea, or trying to obtain a state job, which took months and sometimes years of planning and strategizing. These kinds of initiatives, however, were notably absent from WWF's project reports, thereby perpetuating outsiders' expectations that villagers are passive, require outside intervention for change, and take no initiative themselves. This portrait of essential passivity mirrors the most conservative aspects of resistance models, wherein subaltern agency is mainly a reaction to outside stimulus.

What I am arguing, in contrast, is that by shifting toward engagement models of analysis, we can see that WWF's project plans were transformed through repeated encounters. Bentley's results call into question the idea that as a powerful global institution, WWF could successfully force its will on a local village. Its efforts were continually challenged and sometimes compromised or supported by various government agencies with their own often oblique and sometimes contradictory agendas. WWF had to cooperate with the Nature Reserve Bureau and other state agencies in order to work in the reserve itself. The Bureau staff, in turn, selectively endorsed some of WWF's plans and declined to support others. Meanwhile local residents reacted to the project plans based on their historical understandings and experiences of prior state involvement in their village, comparing WWF's plans to Mao-era terracing campaigns and post-Mao development projects, including cultivating fruit trees, growing cash crops such as watermelon and rubber trees, and increasing their use of pesticides. For Xiao Long residents, the WWF project represented an occasionally perplexing and sometimes threatening set of interventions yet also promised the hope of assistance. Their perspective on the project was no doubt a surprise to WWF staff, particularly to Bentley, who often found himself in a series of confrontations with state officials. His negotiations with them were often prickly and always precarious, since he saw himself as working for villagers in their struggles against state oppression; the complication of such a perspective, however, was that the villagers did not consider themselves objects of state oppression but subjects in search of patrons, or at least relationships with those who had more power and connections than themselves.

Despite the villagers' generalized desire to engage with individuals who might offer hope for the future, it is important to note that they engaged with different levels of energy and ability. Those highly skilled in practicing the art of engagement were often admired, especially those like their old leader, Lao De, who worked to increase the positive and powerful connections that linked their village with the outside world. This was seen as the main way to improve their lives and was perhaps especially important in a village like Xiao Long with few other resources to rally; it is crucial to emphasize that people were more worried about being forgotten than they were concerned about their autonomy.

Resistance models often suggest a very narrow view: that subalterns wish to remove themselves from state forces. Yet for residents of Xiao Long, and many others I met in rural Yunnan, their bigger complaint was the lack of state presence in their village. Many villagers said that higher-level officials were now reluctant to leave their urban comforts and travel to the countryside to share a simple meal with themselves, the *lao bai xing* ("old one hundred names"). As I learned in my fieldwork, some groups actively vie for official acknowledgment of their successes, displays of patriotism, and dedicated enactments of government edicts.[34] There are rewards for being seen as compliant and full of zeal, and I heard widespread misgivings by villagers that their failure to successfully enact WWF's project had led to its termination. Ironically, village "success" or "failure" had little to do with WWF's decision. The funding came to an end and its evaluations were negative not because it failed to turn villagers into excellent agroforesters but mainly because the evaluators saw agroforestry as inappropriate and were convinced that villagers had valuable indigenous knowledge that was ignored.

For the people of Xiao Long, WWF's project represented their most visceral encounter with the environmental winds. They had already experienced nature reserve workers expressing and demonstrating their dissatisfaction with their ways of farming, hunting, and living, which wasn't just deemed backward but in some respects was actually illegal. The repeated presence of foreigners gave them the sense that people beyond China were also interested in their forests and elephants, and provided them with a number of strategies, few of which were deemed successful, but some of which were taken up. I have endeavored to show that we should pay attention to the multiplicity of groups in any global encounter and trace how their interests, projects, and strategies change in relationship to each other. Such a view might help us to move past tendencies to see such encounters as simply a clash of cultures (as a case of resistance) or as a powerful global flow that submerges local sensibilities (as a case of accommodation). By looking at the art of engagement, we can see how forms of desire, reluctance, and strategies of relationship are built by particular histories that inflect the myriad ways that interactions are carried out.

Debates over the project in Xiao Long continued to reverberate long after WWF left the site, fostering new social divisions and ways of political engagement, thereby demonstrating that manifestations of these environmental winds are not merely ephemeral but shape social and physical landscapes in ways that continue to be felt. Only by closely inspecting how globalized social formations are actively negotiated and transformed, at a number of levels, can we gain insight into how global processes are localized or made manifest in particular places. Taking it in another direction, we can see how what we understand as "the global" itself is made and transformed through the art of engagement, through the accumulation of these actions and engagements. The next chapter delves deeper into these dynamics, showing how globalized formations of indigeneity were shaped and made in relation to China's environmental winds.

FOUR

Making an Indigenous Space

OLD ZHANG WAS ONE OF MY FAVORITE PEOPLE to talk to in Xiao Long. He was in his seventies and had lived through dramatic transformations in the world around him. He sometimes shocked his daughter and others by talking so frankly about the past, including the hardships he had experienced. But, as he said, he was not long for this world and he had nothing to lose. One winter day in 2002, Old Zhang squatted over a wild banana stalk, slicing it up with a long, locally forged machete to feed the pigs. He wore a blue knit polyblend cap and sports coat over his tall, gaunt frame, and his machete maintained a rhythmic pattern as he told me of his youth. In the 1930s he had lived in a small upland village. He would periodically travel to a town near the banks of the Mekong River to buy supplies such as salt, lead (for making bullets), and Thai cloth, and during one of these journeys he saw a group of men pointing at him. The men talked among themselves, loud enough to be heard by him and others. Zhang put down his machete and repeated in a whisper what they had said about him decades earlier: "Look, there is one, a *tuzhu ren,* from the mountains." He spoke with both embarrassment and anger, indicating the term's clearly pejorative connotations.

Earlier that month, in an office in the provincial capital of Kunming, the same term, *tuzhu ren,* had been used more positively, albeit tentatively, by Chinese employees at an international environmental organization. For these employees, the term, rarely used these days, was a possible translation of "indigenous people," a social category now used around the globe. What had happened in China to transform this term from a biting pejorative to a positive category that now fosters new ways of understanding politics and social justice for rural people?[1]

In this chapter I explore more of the unintended consequences of the environmental winds in China. One consequence is that the category of indigenous people, along with indigenous rights, is now firmly hitched to plans for conservation and development, as in much of the world. Yet I also show that the forms of indigeneity that emerged in China were not merely the result of accepting a globalized discourse that flowed from outside, that was imposed on or passively diffused into China; rather it is being actively shaped, transformed, indeed "made in China" by a relatively small group of scholars. In turn, new sensibilities associated with the category of indigenous people have begun to reshape the social and natural landscapes of China, as well as the globalized discourse and practices of indigeneity itself.

THE ETHNOGRAPHY OF AN INDIGENOUS SPACE

Nature conservation efforts in China opened up space for the emergence of indigeneity, which was tied to new conceptions of rights, knowledge, and the landscape. In turn, this new vision played a critical role in directly and indirectly shaping the politics of rural livelihood and rights for all people, not just those regarded as indigenous. For example, now when dam projects are proposed in Southwest China (a region with massive hydropower potential), networks of people, both urban and rural, borrow and adapt strategies from indigenous struggles and apply them to a broader social sphere.

Although many people discuss the rise of indigenous rights around the world, which has indeed been a remarkable transformation, less attention has been paid to some of the profound differences in how this phenomenon has worked out in various places. Latin America has been the most frequent model for the powerful social effects of indigenous rights, with many protests, successful land claims, and international coverage of the ongoing Zapatista movement in Mexico and major political reforms, including the election of two indigenous leaders, Alejandro Toledo in Peru in 2001 and Evo Morales in Bolivia in 2005. Yet in Africa and Asia the question of indigeneity remains incipient and fragile. A number of countries, including China, have been officially antagonistic to the question of indigenous rights and have experienced little grassroots engagement. These countries argue that there are indigenous groups in places settled by Europeans (Australia, Canada, New Zealand, the United States, and Latin America in particular), but not in their own territory. Ironically, in China the groups that have most challenged the state, ethnic Tibetans and Muslim Uyghurs, have by and large eschewed the category of indigenous or met with violence when they embraced it. Indigeneity in China (and elsewhere) needs to be understood in relation to China's recent history (Hathaway 2010a; Yeh 2007). In this case, Chinese scholars and scientists smuggled the concept in on the back of the environmental movement, fostering new ways of understanding issues of knowledge, ignorance, rights, entitlements, and citizenship.

In order to grasp the very particular ways that the concept of indigeneity gained traction in China, not just as a term but as a social force in a specific time and place, I employ the concept of "indigenous space."[2] Like the concept of winds, I understand space not as fixed but as dynamic, shaped by history, culture, and human engagements on many fronts. The notion of an indigenous space views indigeneity not as teleological (unfolding toward a predetermined goal) but as a process of emergence shaped by social and historical contingency. An ethnography of indigenous space asks how and why indigeneity becomes relevant or even possible, and how it changes over time in particular places.

THE *NOW* OF INDIGENOUS SPACE (SOME BACKGROUND)

Throughout the world, indigeneity is becoming an increasingly important part of mediating the way development and environmentalism unfold. International organizations in China and elsewhere regularly attempt to determine whether or not their projects include indigenous people; this consideration has now become a requirement. Such linkages started to become significant in the 1980s, when groups like the World Bank faced growing criticism that its many large-scale development projects adversely impacted indigenous groups; in response, they began to build institutional frameworks that altered how projects were accessed and carried out in places with indigenous groups. In the 1990s the World Bank began to change its policy, as Colchester et al. (2001: 84) and others put it, "away from reactive 'mitigation' and damage control towards a proactive 'do good' development [stance that targeted] indigenous peoples." Yet such mandates have caused great consternation to those countries in Asia and Africa that agree with China's stance that all of its people are equally indigenous; hence the challenges raised by indigenous rights are irrelevant to China itself.

The position of the United Nations, however, in response to the stance of China and other states that deny the relevance of indigeneity is that states cannot decide whether or not a group is indigenous; rather it is a matter of a group's self-definition (Xanthaki 2007). Ironically, even though African and Asian leaders disputed the applicability of the indigenous category, it was African and Asian delegates to the UN who created this change of policy away from recognizing the government as the main voice of authority. Therefore this is one way the actions of Asians and Africans, as members of groups not recognized as indigenous in their own country, have successfully changed the terms of the global indigenous debate and the texture of global indigenous movements (see also Hodgson 2011). This is just one good example of many people from many places making the global. The participation of Asians and Africans has changed the way indigeneity itself is defined and acted upon, for it has opened up the category to groups that previously were disenfranchised by their national governments and by other actors that worked against their international recognition as indigenous.

Such a phenomenon has changed the way the World Bank carries out its projects in China. Some of these projects were modified or stopped when they were said to involve indigenous people. More surprisingly, the Asian Development Bank, begun in 1966 as an Asian alternative to the World Bank and expected to eschew indigenous issues, has also started to address them, including in China. This was part of a more general trend whereby other international organizations began to (1) seek out indigenous people for specially tailored projects; (2) advocate for their rights for a variety of reasons, including expanding their opportunities for funding (from funding agencies concerned about environmental or cultural issues); (3) exhibit a growing sense that indigenous rights are important and had been previously neglected or worse; and (4) face internal and external pressures to recognize rights. Increasingly, environmental NGOs in China and elsewhere have tried to collect "indigenous knowledge" and strengthen land rights for those, like Old Zhang, who they would classify as indigenous. Yet although staff at such organizations both generate increasing interest in recognizing such rights and are under greater pressures and incentives to do so, how this will work out in practice is often unclear and is deeply inflected by the particular ethnic formations of the past and present.

Although some Chinese environmental staff were trying to use the term *tuzhu ren* as a way of finding a Chinese translation for the positive connotations and legal rights linked to the international term *indigenous,* Zhang made it clear to me that, at least historically, the term *tuzhu ren* was neither positive nor neutral. After mentioning the term Zhang launched into a brief soliloquy, listing other offensive terms that he had been called in the course of his life, marking each with an extended finger for emphasis. Before the Mao-era language reforms, some of these terms were even more offensive in written form. For example, the Chinese characters for these terms contained ideograms representing "insect-beast radicals," an indicator that such groups were seen as only marginally human (Bulag 2012). Old Zhang concluded, "Now they just call me a *shaoshu minzu* [ethnic minority] or a *nongmin* [peasant]."

Old Zhang had made two things clear: first, finding linguistic equivalents to *indigenous people* in Chinese is not simply a matter of translation; second, and more important, his words reminded me that even if a Chinese term for indigenous people became acceptable to some, the very concept *indigenous* would have to contend with a complex and ongoing history of ethnic formations (see Omi and Winant 1986).[3] Some studies of globalized formations that emerge through social networks and movements have imagined this more as a process of flowing global norms rather than shown us how new social categories and sensibilities move like winds across national boundaries but are also shaped by, and shaping, particular social landscapes.

In this case, it was not that the concept of the indigenous, which is hitched to social networks and forces, just came into a country where everyone is equally Chinese. In fact the government has invested massive physical and ideological resources in mapping and shoring up ethnic differences; for years it has been creating an elaborate system of ethnic categorization that has divided the country into fifty-six "ethnic minority" groups, and this categorization is reinforced in multiple everyday acts. It was also true that the concept of indigeneity was not met with state support, unlike the ways environmentalism was embraced by many state officials and embedded in laws. Instead the official category of indigenous people was deeply opposed by many in Beijing. The concept was little known outside of urban centers, even by people who were themselves potentially indigenous. Indeed my fieldwork showed that it was not rural people like Old Zhang who proclaimed themselves indigenous but mainly Chinese experts who used this transnational concept in an effort to reshape notions of ethnicity, citizenship, and rights.

In other countries, debates about indigenous rights often circulate around legal treaties or questions of which groups have more rights due to historical precedence, neither of which have gained any real ground in China. Rather, in Yunnan the issue of indigeneity blew in with the winds of environmentalism, and it was in that context that it was discussed. Before trying to understand how environmentalism articulated with indigeneity in China, I explore some of the ethnic formations that have informed Old Zhang's worldview.

CHINA'S ETHNIC FORMATIONS

China is a place quite unlike other countries with a long-standing popular interest in indigenous groups. For instance, among citizens in Brazil (Ramos 1998), India (Baviskar 2005), and the United States (Redford 1991), there are complex notions of indigenous people, including many derogatory associations, but also with a widely held romantic sentiment about "Indians" and "tribals" as "noble savages," evoking positive connotations of innocence and ecological stewardship. In Yunnan in the 1990s there was little sense of the noble savage.[4] Although some cultural groups in China, especially those living far from urban centers, might have been seen, according to a Western framework, as "traditional peoples," Chinese media and urban dwellers have tended to describe such groups as backward (*luohuo*), lacking in culture (*meiyou wenhua*), and dirty (*zangde*). Many anthropologists have written insightfully about the ways that ethnic minority groups in China are an object of curiosity, touristic performance, and derision (Blum 2000; Harrell 1995; Litzinger 2000; Mueggler 2002; Schein 2000; Walsh 2005).

While there are parallels with how indigeneity is signified in Western countries, there are also significant differences. For example, in the Americas relatively few groups have a written language, and whether a group does or does not have one is seen as relatively inconsequential. Instead indigenous language ability is emphasized, and those groups that have bilingual representatives provide positive evidence of indigenous authenticity (Ramos 1998). In China, however, the Han, who pride themselves on having had a written language for thousands of years, tend to stigmatize as grossly deficient and uncivilized the many ethnic groups that lack a written language (Hansen 1999; Ramsey 1987).[5] Chinese descriptions often foreground an ethnic minority group's possession or lack of a written language, and groups without writing are seen to have a low level of social development (Zhou and Sun 2004).

Since the end of imperial China in 1911, leaders have attempted various strategies to classify the people living on Chinese territory, in their attempts to acknowledge and manage ethnic difference as part of different nation-building projects. In the early 1900s officials declared that China had five different groups: Han, Manchu, Mongol, Tibetan, and Muslim.[6] Many textbooks from the time discuss non-Han people as "foreigners" (*wai*) or enemies (*di;* Baranovitch 2010). Soon after the inception of the People's Republic of China in 1949, the world's largest ethnological survey was carried out in an attempt to codify a bewildering diversity of languages and lifeways. At one point, over four hundred ethnic designations were considered in Yunnan alone, but only twenty-five were accepted: currently Beijing recognizes fifty-six ethnic groups across the whole of China. Of these, the national majority Han Chinese are said to be around 90 percent of China's population, and other groups, such as ethnic Tibetans, Dai, and Yi are called *shaoshu minzu* (literally, "minority nationalities") and are colloquially referred to as *minzu* (nationalities). It should be remembered that as China is quite populous, some groups, such as the Yi, little known outside of China and constituting less than 1 percent of China's population, are nonetheless more numerous than the Irish.

China's ethnological surveys were influenced by long-standing notions of ethnic difference and particular social theories that came to China as part of transnational socialism. One of these theories originated with the American lawyer and ethnologist Lewis Henry Morgan, who elaborated his notion of cultural development in the late nineteenth century. Morgan's studies were discovered by Friedrich Engels, then shared with Karl Marx, who read them with great interest and rewrote his *Communist Manifesto* and other important documents based on Morgan's theory of history and social development. In turn, Marx's reiterations informed the Soviet approach to classifying and governing the large number of ethnic groups that made up the USSR (Grant 1995). These ideas were later brought to China, where they were subsequently modified again. During China's ethnological surveys, each ethnic group was placed into a single evolutionary stage: primitive communist, slavery, feudalist, or capitalist. Despite the fact that the Chinese Communist Party saw capitalism as the world's main problem, only the Han majority was said to have achieved this high level of social development, which was, in fact, a source of Han pride. Conversely, there was little to celebrate about those groups labeled as primitive communists. PRC historians declared that the Han people passed through that stage over three thousand years ago, so that the "primitive communists" discovered in the 1950s were seen as "living fossils" (Dirlik 1996), a glimpse into what Han society looked like ages ago. Given this, the state's main imperative was to help each minority group make "great leaps forward" (jumping from a society that was primitive communist or slave-based directly into a socialist stage). Many reports from the 1960s and 1970s celebrated such transformative leaps, implying that most of the ethnic groups had now become equally socialist.

By the 1990s, however, this narrative was either largely forgotten or ignored, so that different minority groups were often described as being locked into the evolutionary stage in which they were "discovered" in the 1950s (Hansen 1999). Yunnan is cultivating "ethnic tourism," promising to bring tourists to witness groups that exist in a "previous era." Although some Chinese academics have begun challenging these theories of social evolution, such theories are still widely accepted in China as a fact (Tong 1989; Litzinger 2000). Indeed I often heard individuals in rural and urban China recite the five stages of society as they had learned them in school.

Even now, when the notion of evolutionary stages is rarely employed, the concept of backwardness (*luohuo*, literally "behind the times") remains key in discussions of ethnic minority groups and "peasants." References to minority groups are still often based on a continuum, with the Han almost unanimously conceived of as the most "advanced" and other groups regarded as relatively advanced or relatively backward—still implying a powerful sense of social evolution and development.

To be fair, the notion of backwardness has wide usage, including indexing China's place in the global ecumene (compared to countries in Europe and Africa), as well as describing people and places within China. Usually it is not applied to individuals so much as to groups of people. There is a way in which Han ethnic pride can stimulate its use, and I have heard poor Han citizens describe their ethnic minority neighbors, even if the neighbors are financially better off, as backward. Usually this claim is in reference to ethnic minorities not following desired family-planning goals or not emphasizing their children's educational careers—and even if, empirically, these assertions have no merit. In this way, the sense of Han pride is similar to that of white people in the United States, as insightfully analyzed by critical race scholar David Roediger in his book *The Wages of Whiteness* (1999).

Recently some minzu intellectuals have challenged popular and official characterizations of their group as backward and have asserted their group's positive contributions to China. Their examples tend to resonate with state-recognized achievements, such as how their group supported Mao's Red Army in the 1930s and 1940s or how they have used scientific principles since ancient times to carry out good family planning (Litzinger 2000). In other arenas, minzu musicians confront Han-dominated notions of cultural citizenship through traditional ethnic performances (Rees 2000), and others, such as ethnic rock stars, reappropriate derogatory ethnic terms such as *Lolo*, once used by Han to describe non-Han "barbarians" (Baranovitch 2001). These efforts attest to the wide-ranging strategies deployed by various minzu intellectuals and artists to reimagine the place of non-Han Chinese within the nation. Unlike in North and South America, however, there have not been pan-ethnic movements in China. Rather boosterism (whereby intellectuals or officials advocate for their particular ethnic group) is much more common. As far as I know, minzu leaders within China have not rallied under the rubric of indigenous peoples.[7]

As I will show, within these social hierarchies and frameworks in China, one of the ironies is that it has been easier to position as indigenous those groups that are seen as "relatively advanced" in China than those groups regarded as "relatively backward," even if it is the latter who more closely correspond to Western expectations of indigenes, those who ostensibly possess the attributes that fit easily into what Tania Li (2000) calls the "tribal slot."

Official and popular conceptions of ethnicity in China are entangled with the state's efforts to classify and regulate it (Blum 2000; Harrell 1995; Mullaney 2004; Baranovitch 2001). There is little sense in China that these groups will ever be given political autonomy as "nations," and there are almost no debates over the nature of indigenous sovereignty or "self-determination," as there are throughout the Americas. During the first decade of the PRC, there were some promises and negotiations around self-determination, but by the late 1950s this issue had been largely settled, and there were few opportunities to pursue it in any substantive way in the official political sphere. There have been few substantial challenges regarding names or designations since these groups were solidified in the late 1950s; after 1979 the state declared that no new minzu would be recognized.[8] Today ethnic identification in China is mandatory, naturalized, and singular (not multiple, hyphenated, or fractionated, as elsewhere around the world). Government-issued ID cards declare one's minzu status, and I have noticed that lists of students or employees often contain people's names, dates of birth, and gender and minzu designations. This system is firmly in place, and while there have been some notable attempts to improve depictions of some ethnic groups in history books, for example, there has not been a widespread grassroots indigenous movement in China.

THE ARTICULATION OF ENVIRONMENTALISM AND AN INDIGENOUS SPACE

Beginning in the 1980s indigenous social movements began to significantly reconfigure the laws, politics, and everyday life in many countries (Brysk 2000; Friedman 1999; Kingsbury 1998; Warren 1998) and gained support within a number of international arenas, such as the United Nations (Niezen 2003; Muehlebach 2001). Global monetary organizations and NGOs were also strongly influenced by the international indigenous movement, and in the 1990s indigenous advocacy groups pressured organizations like the World Bank and the World Wildlife Fund to create special guidelines for projects involving indigenous peoples.

Yet indigenous status is not always readily apparent or easily achieved, and to "become indigenous"—participants in a political category—groups must be "positioned" within the transnational solidarities, rights, and participation of a dynamic social movement (Hodgson 2011).[9] In other words, popular beliefs notwithstanding, indigeneity is not a natural or primary category but one that emerges in articulation with social and political ideas, laws, and institutions (Jung 2008; Tsing 2003). In the year 2000, although the World Bank labeled some groups in China as indigenous,[10] I found little evidence that rural groups in Yunnan identified themselves that way.

Rather I contend that a space of indigeneity emerged primarily through the efforts of a small cohort of Chinese, mostly Han, experts, including some who worked as staff and consultants for international environmental NGOs, as well as for a broad range of Chinese-based institutions and NGOs. This cohort used the English term *indigenous peoples* along with a linked set of conceptual frames such as "indigenous knowledge" and "sacred lands" (Sturgeon 2007). Even though some foreign staff at international NGOs in Yunnan saw some of the ethnic groups they worked with as indigenous, they were not advocates for indigenous rights per se. Staff at these organizations did not always agree on which groups should be considered indigenous. For example, many staff at The Nature Conservancy (TNC) worked hard to establish an indigenous space for ethnic Tibetans but virtually ignored many neighboring minzu groups. Of all the groups in the region, Tibetans were the most politically visible, and, as some TNC staff feared, working with Tibetans would arouse the displeasure of government officials. While TNC (2003) has endeavored to position itself as science-driven and apolitical, its stance could easily be regarded as politically contentious. It was not only officials who challenged TNC's engagements with minzu groups. Some Chinese experts themselves worried that if TNC tried to expand Tibetan rights, it might attract the accusation of working toward "splittism" (*fen lie*), an official government term for the condition of breaking up the unity of China, and especially used for Tibet and Taiwan. Several of these Chinese experts, some of whom were employed by TNC, began to challenge TNC's work and positioning, especially when TNC seemed to hamper the rights of non-Tibetan minzu.

Despite such complications, it was primarily the active presence of environmental organizations in China in the early 1990s that provided increasing opportunities for indigeneity to emerge. As we have seen, the environmental winds were blowing more strongly through Yunnan than other provinces, and conservation projects were proliferating but were also starting to be judged in new ways. Previously conservation organizations had seen rural inhabitants as peasants, not only employing that term but also treating them as such, that is, as people to be reeducated, removed, and reformed, as objects to be trained in scientific methods. Yet by the 1990s a new understanding of people and the environment was emerging. Project staff from many of the NGOs were compelled by a growing conviction in international circles that indigenous peoples have a unique relationship to the environment, possess a broad repertoire of indigenous knowledge, follow livelihood practices that might sustain the environment, and should be entitled to greater land rights. Although environmental NGOs are not necessarily committed indigenous advocates (Chapin 2004), some NGOs began to offer more support to groups regarded as indigenous, either directly or by lobbying states to award more rights, thus creating incentives for groups to identify as indigenous. Yet this category was not, intentionally or unintentionally, just smuggled in full form from abroad. Instead Chinese experts and others labored to gain traction for this new social category, negotiating between the language, connections, and resources of international environmentalism and existing Chinese ethnic formations. They used their skills, research, and advocacy to give political and social meaning to the question of the indigenous in China, and did so using the language of environmentalism. In what follows, I examine the lives of several of these experts, some of whom have been mentioned earlier in the book. Their efforts provide insights into the difficulties and diverse strategies behind the creation of an indigenous space.

PROPONENTS OF AN INDIGENOUS SPACE: CHINESE EXPERTS

Pei Shengji, Xu Jianchu, and Yu Xiaogang created and reworked the potentials and meanings of an indigenous space in Yunnan. Although their lives overlapped at various points, their sensibilities, approaches, and career paths have differed widely, and their aims and strategies have changed over time.[11] Pei Shengji, born in Sichuan in 1938, was one of the first in Yunnan to connect with international indigenous discourses, conducting his work in the style of the cosmopolitan gentleman scientist. Xu Jianchu, Pei's student, born in 1964 near Shanghai, also continued a scholarly tradition but did so in relation to Yunnan's booming NGO activity, building a vibrant network of scholars and domestic and international NGOs. Yu Xiaogang, born in 1951, combined work with his own NGO, international NGOs, and direct activism. Of the three, Yu was the only one who actively confronted development projects by environmental NGOs and the Chinese state. In places such as Latin America, the most vocal advocates for indigenous rights often identify as indigenous themselves, whereas Pei, Xu, and Yu all identify as Han Chinese, the ethnic majority. Their work has been far reaching, not only challenging a widespread understanding that China is a land of cultural homogeneity (many foreigners imagine China as a country inhabited only by essentially similar "Chinese") but also changing the everyday dynamics of governance, conservation, and ethnicity in rural Yunnan. They influenced many peers, dozens of whom consult for foreign and domestic organizations and agencies. They gained traction for the concept of the indigenous in different ways and inspired foreign organizations and researchers to investigate indigenous-related questions and issues.

Finding Sacred Lands in China: The Work of Pei Shengji

When I first spent a year in Yunnan, in 1995, many of my Chinese colleagues worked for international conservation NGOs, and there was almost no discussion of "sacred lands" or "sacred forests." When I returned for fieldwork from 2000 to 2002, dozens of reports on the sacred lands of Yunnan were available in Chinese and English (e.g., Yang et al. 2004; Liang et al. 2009; Long and Zhou 2001). Two of the world's wealthiest environmental organizations, The Nature Conservancy and Conservation International, were carrying out major projects to document and protect sacred lands, working in partnership with government officials to establish park-like protection for these areas (Litzinger 2006). Almost all of these people referenced their debt to ethnobotanist Pei Shengji.

Pei, now known as the father of Chinese ethnobotany—that is, the study of the relationship between people and plants—has significantly influenced how scholars and others know and think about human-nature relationships in Yunnan. He himself played a key role in the transnational work of connecting Yunnan with international interest in biological diversity and its intersection with cultural diversity. During the 1980s Pei was based at the Xishuangbanna Tropical Botanical Garden, very close to where Old Zhang lived, near the Mekong River and the borders of Vietnam, Laos, and Myanmar. The Gardens were just being rebuilt and restaffed after great difficulties during the Cultural Revolution. At the botanical garden, Pei read about the concept of sacred forests, which initially emerged from studies in India and Africa (Clarke 1976; Gadgil and Vartak 1976). By the time Pei read about it, the concept had already been expanded and taken up by environmental and indigenous rights organizations, especially as they began to seek common cause, beginning in places like the Amazon (Conklin and Graham 1995). Sacred forests became an example of "indigenous conservation," a concept that helped initiate new ways of organizing and designing environmental projects. It was embraced and promoted by organizations such as the United Nations Educational, Scientific, and Cultural Organization (UNESCO) and other influential actors (Hay-Edie 2004). The concept of indigenous conservation was one of the many factors that mitigated against the long dominant approach to conservation described as one of "fences and fines," which worked to criminalize and vilify local peoples.

In 1985 Pei published his first English-language essay about sacred forests in China, which appeared in the anthology *Cultural Values and Human Ecology in Southeast Asia*. His essay tied him closer to Southeast Asian networks, culminating in his involvement with the Southeast Asian Universities Agroecosystems Network, which connected him and his students to peers in Indonesia and the Philippines. At the botanical garden, his peers had been focused mainly on finding medicinal and utilitarian plants to enhance national security and well-being. For instance, they discovered that China possessed a species that could substitute for a valuable medicinal herb that had long been imported from India; another breakthrough discovery was a tree sap that could provide excellent antifreeze qualities that would allow Chinese military planes to fly at higher altitudes and in colder climates. With few exceptions, they were far less interested in how minzu protected the forest or enhanced biodiversity. As far as I know, Pei's essay was the first to apply the emerging international framework of indigeneity to Yunnan. This framework included the concepts of sacred lands, indigenous knowledge, and indigenous rights. His reports were among the first English descriptions of minzu that significantly deviated from standard Han modes of depiction, where, at best, minzu might be successfully tutored in the principles of rationality and science. Instead he framed their knowledge and practices as hopeful models for "sustainable development" and the "conservation of biodiversity."

His initial studies were on the Dai minority and their Holy Hills or Dragon Hills (*Long Shan*). Pei's studies made the Dai practices visible, valorized them, and attracted the notice of social and natural scientists from China and abroad, who in turn came to study the Dai and write their own reports. Each positive report further solidified the Dai's conservationist reputation. I heard from one colleague that for a while in the early 1990s, a number of foreigners came to Banna to study Dai home gardens, stimulated by Pei's writings in English. For a while, home gardens, rich in biological diversity, seemed a model form of ecological and social development, especially in Asia (Saint-Pierre 1991). Many foreigners just passed through Banna quickly, but Pei's work also pulled in some who did research, stayed in Yunnan, and helped shape the way environmental activities were worked out in the region.

One example of Pei's influence is Ulrich Apel, a German forestry expert working for GTZ. In Germany Apel discovered Pei's work and found it quite exciting. It countered his impression, shared by many of his peers in Europe, that China was an ecological wasteland and a cultural monoculture. As Apel said, "At that time in Germany I had no idea there was rain forest in China or that tribes lived there." In turn, Apel traveled to Dai villages to conduct his PhD fieldwork in forestry under the guidance of staff at the botanical garden. He documented how the diversity of trees and shrubs was changing in Dai sacred forests. His dissertation, written in German, was translated into English by an inspired colleague, and this translation was later published by GTZ's China office. As with Pei's studies, Apel's work had far greater reach after it was translated into English. His study circulated within Yunnan and was read in tandem with the studies Pei had done almost two decades earlier, which had been its initial inspiration. Apel became a project evaluator for international conservation projects in China and Vietnam. Like the EU experts who came to evaluate the WWF's project in 1995, Apel criticized existing or planned projects that regarded ethnic minorities as ignorant and limited their rights to local resources, and he thereby propagated the new sensibilities of an indigenous space. Their work contributed to the sense that certain ethnic minority groups had religious connections to forests, which, rather than being dismissed as lingering forms of superstition, should be scientifically visible and viewed as ecologically valuable.

Like Apel, and unlike most anthropologists and indigenous advocates, Pei used less of the language of romantic prose and more of the language of science to document these sacred forests. He used Latin names and species lists to show the range of plant species present and compared diversity levels between sacred and unprotected forests (Pei et al. 1995). He conducted botanical transects of the Dragon Hills, and by combining field data and carefully generated statistics he was able to convince a number of domestic and international conservationists of these forests' ecological importance and thus the positive environmental role played by the Dai.

Of all the ethnic groups in the region, why were the Dai among the first and best studied minzu highlighted in international forums on an indigenous space? This situation was unlike other places in the world where groups that adhere more closely to international expectations of indigeneity, like the Maasai or the San, are more likely to gain attention, respect for their knowledge, and privileged rights as indigenous peoples than their less "tribelike" neighbors. Ironically, groups such as the Maasai and San have long struggled domestically, as they are allocated a low social status, their lands are targeted for expropriation, and their bodies are targeted for "social improvement" (Hodgson 2002; Sylvain 2002).

In Banna, in fact, the Dai were seen as the least tribe-like of all the regional minzu. It was easier for officials, scholars, and experts to accept that the Dai, compared to their upland neighbors, possessed valuable knowledge. Indeed a number of local and regional officials were Dai themselves. Uplanders were seen as nomadic groups, without a system of writing, who maintained primitive beliefs (beliefs often dismissed as "animism" and written off as an incoherent set of superstitious ideas) and carried out primitive slash-and-burn agriculture—a practice that, by the late 1980s, was coming under criticism as environmentally destructive and was often outlawed. The lowland Dai, on the other hand, had been rulers of a large Buddhist kingdom for centuries before the region was brought under Beijing's ambit in the 1950s.[12] They are a relatively well-off group with substantial fields of high-yielding paddy rice agriculture, deeply involved in commercial trade; they practice Buddhism (one of the five religions officially permitted by the Chinese state) and have their own writing system. The combination of these attributes fits Han Chinese notions that the Dai are a "relatively advanced" (*bijiao fada*) minzu, higher than their minority neighbors but lower than the Han.

Pei's diverse efforts, which helped foster international interest in Yunnan Province, are an excellent example of the kinds of transnational work that was required to create the linkages that would articulate an indigenous space. In 1990 Pei was a major organizer for the second International Ethnobiology Meeting, likely the largest international conference yet held in Yunnan Province and one that brought together hundreds of foreign researchers. Another organizer for this conference was anthropologist Darrell Posey, who, before his untimely death, was one of the world's most influential advocates for connecting indigenous rights and environmentalism, starting with his work in the Amazon (Hern 2011). At the Kunming meetings, Pei, Posey, and Xu Jianchu lobbied to produce the "Kunming Action Plan," which stressed indigenous intellectual property rights and new ethical protocols for working with indigenous peoples; these were powerful statements and agreements unprecedented in China. Posey and Pei helped forge these initial linkages, and the process of carrying them out in and beyond China relied on the work of Pei, Xu, and many others.

This plan also initiated the Global Coalition for Biological and Cultural Diversity to "encourage the permanent and meaningful dialogue between indigenous peoples, scientists, and environmentalists in order to develop a unified strategy to defend the biological and cultural diversity of planet Earth" (International Society of Ethnobiology 2009). The meeting itself not only stimulated novel frameworks and practices among Chinese scientists but also jump-started programs that reconfigured globalized debate around ethics, indigenous rights, and nature conservation, thus directly contributing to "making the global." By bringing together hundreds of scholars and staff from environmental and cultural rights–based NGOs, the conference helped promote international awareness and interest in Yunnan and expand Pei's professional networks.

Pei's networks and publications (1985, 1993; Pei et al. 1997) were key to attracting the attention of environmentalists and others interested in Yunnan's cultural diversity. His international reputation helped increase his status within China, and officials increasingly asked him to play influential roles in steering government projects, especially those engaging foreign organizations. His career trajectory brought him from the Xishuangbanna Tropical Botanical Garden, situated in a relatively backwater location in rural Banna, to bustling Kunming, a major city. He established a department of ethnobotany at the Kunming Institute of Botany, part of China's prestigious Academy of Sciences, then moved to Kathmandu after he was invited to head the International Centre for Integrated Mountain Development (ICIMOD), an influential think tank for the Himalayan region. Prior to Pei's leadership, ICIMOD was focused mainly on Nepal and India and had relatively little involvement with China. At ICIMOD he helped pull Yunnan further into the conversations and debates around the Himalayas, gained international attention, and expanded the networks of Yunnan experts. Returning to Kunming years later, he helped reconfigure Chinese ethnobotany, which was originally designed to harness the wealth of minzu botanical knowledge for national goals. Chinese ethnobotany has now, as in Mexico and some other countries, expanded its focus from exploitation to a greater recognition of indigenous knowledge, sacred lands, and intellectual property rights (Hayden 2003).

Sacred Lands and Feudal Superstition: The Uneven Work of Displacement

Although Pei's work on sacred lands provided new ways of understanding Yunnan's social landscape, it wasn't easy to displace deeply embedded perceptions that had become part of the social landscape. In particular, his use of the terms *sacred lands* and *sacred forests* had to compete with older notions of superstition (*mixin*).[13] This challenge paralleled attempts by Chinese experts who were trying to find terms to translate the positive connotations of *indigenous people*, given that the most popular translation of that term, *tuzhu ren*, was already contaminated, as Old Zhang showed us. Although there is rarely any mention of *mixin* in the texts that document or celebrate the existence of China's "sacred forests," this burden of language and the history of its social affect is in fact one of the biggest obstacles to the spread of these new concepts in China beyond a relatively small group of experts.

This situation became clear to me during my fieldwork in Northwest Yunnan in 2002, when I talked with a nature reserve official. We discussed a forest grove that would later be referred to as a sacred forest in UNESCO documents. I asked this official, Hu Lihua, where I could see examples of older forests, preferably some remnants of the "original" forests. The nearby slopes were full of scrubby pines. Looking at a large plaster model of the surrounding watershed, Hu pointed at one place and said, "Around here, they still have big trees. Big oaks." I asked why that patch remained and not others. Hu looked down, considered his response, and somewhat sheepishly answered, "*Yinwei tamen de fengjian mixin* [Because of their feudal superstition]." *Fengjian mixin*, a Marxist historiographic term, which, like many such terms, came from Japanese Marxists, was a key concept used throughout the Mao era to criticize past beliefs as unscientific and irrational.[14] The term was especially consequential during the Cultural Revolution, when it motivated violence against people (such as shamans) and objects (such as temples or statues) seen as emblematic of superstition (Dirlik 1978; Hathaway 2010a). Unlike terms such as "traditional" (*chuangtong*), which can have a positive valence and suggest something worth saving, *mixin* is always problematic, describing "outdated beliefs." It is used, for example, by contemporary Chinese teenagers to criticize their parents' old-fashioned notions.[15]

As Hu saw it, the continued presence of these old trees was evidence that feudal superstitions had survived the turmoil of the past. Although he was only a child during the Cultural Revolution, for him, the trees still represented a continuing struggle between superstition and science. These old trees revealed the state's lack of success, and indeed his own lack of success, in fostering more scientific understandings of the world. In fact he saw his job as teaching peasants about scientific management. In some ways, it was remarkable that this forest patch remained; it shows that locals played an active role in protecting these trees while most of the surrounding lands were deforested. Especially from the 1950s to the 1970s, many of the older trees were cut down as villagers carried out campaigns during and after the Great Leap Forward to convert forests into grain fields and trees into fuel for smelting steel. Some "antisuperstition" campaigns intentionally targeted the same groves of trees that would later be seen as sacred forests, and many were severely depleted or eliminated. Hu's awkwardness cannot be explained as merely Han condescension toward minzu. He himself identified as Naxi minzu, the same ethnic group as the communities that protected the forest groves. Nor can his reaction be explained as an example of a deep antagonism between the state and ethnic minorities. Indeed at the time of the Cultural Revolution, Naxi leaders were regarded as stronger supporters of the Chinese Communist Party than the Han, and Naxi enthusiastically carried out government campaigns (White 1997).

Reports by the nature reserve staff mentioned this patch of old-growth forest, although not as an example of sacred forests. In spite of Hu's awkwardness about the history of this patch of forest, other reserve staff later touted it as a valuable example of "untouched nature," even though it remained not as the result of human neglect but as the result of consistent human protection.[16] There is some general sense of sacred lands in China, but these lands typically include only well-known mountains like Emei Shan that have long been part of Buddhist sacred topographies seen as belonging to all national citizens. Concepts that link notions of ethnically specific sacred lands to indigeneity and political rights have only started to take hold among a variety of urban-based and international organizations in China. The notion of indigenous sacred lands still has little traction in rural areas, whereas the idea of superstition remains strong there and elsewhere (Chao 1996; Hathaway 2010a). This is very different from the practice in some countries, such as Australia, Canada, and the United States, where the notion of sacred lands has gathered political and legal force and forms the basis for successful indigenous land claims (Tsing 2007). Following the paths of these concepts alerts us to how globalized formations, and the constellation of social and political concepts that are part of their frameworks, imbricate quite differently among, as well as within, places, based in large part on their sociocultural pasts.

From Slash and Burn to Swidden: The Work of Xu Jianchu

Like Pei Shengji, Xu Jianchu worked as an academic and consultant. His career progressed from earning a degree in horticulture at a well-known university in northern China to becoming one of China's key cultural brokers between international organizations and Chinese institutions. He studied under Pei in Xishuangbanna, earning a master's degree for his study of how one upland group created indigenous management systems to manage rattan vines, a non–timber forest product. In 1992 he earned a second graduate degree in an English-speaking university in the Philippines and creatively employed his cultural fluency in international environmentalism. He played a major role in shifting environmental discourses in Yunnan away from a view that peasants destroyed nature due to ignorance and backward farming practices, toward a more serious consideration of indigenous knowledge. He published widely in Chinese and English (Xu 1990; Xu and Mikesell 2003; Xu, Li, et al. 1995; Xu, Pei, et al. 1995; Xu et al. 1997, 1999), worked as a consultant for many international environmental organizations in Yunnan, established a highly successful NGO, the Center for Biodiversity and Indigenous Knowledge (CBIK), and orchestrated major international conferences that showcased his work and that of his colleagues. By drawing on the rubric of indigenous knowledge, Xu challenged dominant environmental discourses, most notably negative characterizations of slash-and-burn agriculture. As described in chapter 2, in the late 1980s many Chinese officials and foreign environmentalists alike regarded slash and burn as the biggest threat to the world's tropical forests.[17] This sense of threat resulted in projects to relocate villages out of areas deemed ecologically valuable. Officials often banned slash and burn without leaving villagers any viable alternatives (Yin 2001).

During the 1990s some environmentalists were hesitant to use the phrase *slash and burn*, as it became seen as unnecessarily derogatory rather than merely descriptive (O'Brien 2002). As part of this changing sensibility, in 1955 some scholars began to use *swidden*, a term from Old English, as more neutral (Eckwall 1955). As elsewhere in the book, I use the two terms according to their original use; they are not synonyms but different concepts. As I show, the concept of swidden agriculture as a neutral or positive practice did not automatically gain acceptance in China. Rather it gained a foothold through the actions of Xu and several others, most notably Yin Shaoting.[18] Xu created a conceptual distinction between slash and burn, as practiced by cash-crop farmers, and swidden, as practiced by indigenous peoples. His work dislodged existing understandings and forged new views in three main ways, creatively inventing a Chinese notion of swidden. First, he reexamined the scientific reports that blamed slash-and-burn agriculture as a major source of soil erosion. These reports had compared erosion from slash and burn with erosion from undisturbed forest plots—a comparison that made slash and burn seem quite detrimental and was used as a compelling argument to ban the practice. Xu, however, was among the first in China to carry out new studies that compared erosion from swidden with erosion from other agricultural practices, such as the establishment of rubber and tea plantations. He argued that swidden must always be compared to alternative land use practices. This slight reframing had dramatic consequences: it showed that, compared to other land uses, swidden was less, not more, erosive.

Second, whereas most scientists regarded slash and burn as a backward practice that persisted despite state disapproval, Xu was one of the first to reveal that in fact it had been advocated by state officials. He showed that, in the 1960s and 1970s, the state had encouraged farmers to slash and burn in order to increase grain and to remove forests in order to create state-run rubber and tea plantations. He highlighted how state-led forces played a greater role in upland deforestation than independent farmer-led agriculture. He challenged conventional wisdom that saw upland farmers in spatial and historical isolation, as living on the margins or outside of state rule. In a conventional view, peasants' slash and burn was said to defy state regulations,[19] but Xu showed that their actions were mandated by government-led campaigns for grain production and plantations. His work helped foster the emergence of an indigenous space in China and transform some conservationists' understandings of the relationship between local peoples, the state, and the environment. Prior to this, many expatriate conservationists reinforced Han prejudices of minzu (e.g., MacKinnon et al. 1996).[20] Later many conservationists saw minzu as holders of valuable knowledge (Anderson et al. 2005; López-Pujol et al. 2011). Before, most conservationists saw themselves as allies of the state, working together to protect forests from encroaching peasants. Later, a number of conservationists questioned this alliance, seeing the state as the main perpetrator of destruction and thinking it better to encourage local management.

Third, Xu appealed to conservationists by showing that swidden agriculture led to much higher rates of biodiversity than other land-use alternatives. Previously social scientists who were sympathetic to swidden often remarked that swidden plots appeared diverse and "like a jungle" (Geertz 1963: 19). Such statements about its visual appearance were not of much value to scientists, who wished to specify and quantify all of the species found in swidden plots—similar to what Pei had done with the Holy Hills. Xu, however, not only created detailed species inventories but also performed statistical analysis, an additional layer of labor that allowed him to connect swidden more tightly with concepts used in conservation biology such as "richness," "biodiversity indices," and "evenness." His reports were respected by practicing biologists and conservationists. By using quantitative claims in novel ways, he invented new and convincing arguments on behalf of Yunnan's upland farmers, challenging environmentalists to look for biodiversity not only in "the wild" but also in farmers' fields. As he and his colleagues declared, it is important to recognize that "swidden cultivation, rather than being a threat to tropical biodiversity, may be the most appropriate and culturally suitable means for preserving biodiversity" (Xu et al. 1999: 131). His work encouraged officials to look beyond farmers as the main sources of erosion and to examine state-led plantations as possible culprits. Eloquent and cosmopolitan, Xu spoke in the idiom of science at a time when science in China was quickly gaining in respect, and he skillfully negotiated his role as an intellectual and scientist.

Although I emphasize Xu's efforts, colleagues with similar sensibilities were also helping to shift the understandings of the environment, the state, and rural citizens. It is important to realize that Pei and Xu helped create the institutional space to fund, support, and legitimize this work. Such studies would have been quite difficult into the early 1980s, before Pei started to foster these perspectives. His colleagues also took up the mantle to compare the levels of bird diversity found in areas that used traditional swidden versus more conventional modern planting methods, showing that the former led to greater levels of diversity (Wang and Young 2003). There are now dozens of scientists who carry out studies in Yunnan that reconceptualize traditional beliefs not as superstition to be reformed but as an important basis for nature conservation (Chen et al. 1992; Liu et al. 2002; Yang 2007; Luo 2008). Others too have joined in the critique of what have been long-valorized forms of agriculture, such as chemical-intensive paddy rice, now described as Han-imposed, ecologically destructive, or inappropriate (Li 2001). They join others writings about the Mao era as a time of environmental degradation (which resulted from, for example, the plowing up of grasslands or the filling in of wetlands that removed natural ecosystems and yet failed as agricultural fields), but do so in a particular way that highlights the sustainability of minzu practices (Yang 1998 in Shapiro 2001; Hansen 2005). In so doing, they are not only promoting elements of an indigenous space; they are in fact making it.

In China indigeneity has emerged and continues to do so through environmental discourses. Elsewhere in the world, such as in the United States, Canada, and Australia, arguments about indigenous rights have been primarily framed in terms of morality or historical precedence and debated in the courtroom. These frames, however, have not been important in the Chinese context. Rather the concept of indigenous became important through concerns about sustainability and the efforts of Chinese experts concerned with promoting environmentally friendlier practices and policies. Writings on indigeneity and ecology are not just academic; people like Xu targeted conservationists and regional officials—people who were in charge of making important decisions in determining land-use rights for rural minzu. In places like southern Yunnan, the first major hub for domestic and transnational nature conservation in the province, these experts entered into a social landscape with a long-standing division in political power and representation of lowland and upland groups, where uplanders were often lacking in political power and economic wealth. As mentioned earlier, among the various ethnic minority groups in Banna, the ethnic Dai who lived in the lowlands were the main focus of Pei's research and enjoyed a relatively high social status before his writings. When foreign conservationists and regional officials began implementing projects in the 1980s, they generally viewed the Dai positively, as possessors of ecological knowledge (referencing Pei's studies, which had been translated into English, on the Dai's sacred forests and home gardens), and they typically adhered to Chinese ethnic hierarchies that derided uplanders as ignorant peasants who were destroying the forest (Cheung and MacKinnon 1991). This dichotomous view is evident in the WWF's project reports, as well as in a film made by MacKinnon, "The Story of Xishuangbanna" (1991). When the WWF began its projects in Xishuangbanna, there was no sense whatsoever that these uplanders might one day be seen as indigenous people with internationally recognized rights.

When Xu carried out his studies and published his charts, he was therefore among the first to present statistical evidence that uplanders created, and did not threaten, the biodiverse landscapes that conservationists and officials wanted to protect. A few Chinese anthropologists such as Yin Shaoting had written positive accounts of uplanders' livelihood practices, but these accounts had little influence among conservationists and scientists. Xu was the first to reframe the discussion within Chinese conservation by invoking concepts linked to indigeneity. As one expatriate conservationist working in Yunnan told me, anthropological accounts of uplanders "might be interesting to read if they were even available in English, but they are unsystematic and unreliable. But Xu Jianchu's charts were clear and convincing—they make a strong case that indigenous knowledge plays a big role in preserving the environment, even in China."

FIGURE 12. Dai sacred landscape. Advocates for an indigenous space conducted studies, carried out quantitative research, and created new images like this to argue for the existence of sacred lands in China. Artwork by Yang Jiankun, 2005.

FIGURE 13. Hani sacred landscape. This image was one of the first to depict upland ethnic minorities in China as having sacred forests. Artwork by Yang Jiankun, 2005.

Xu wrote for Sinophone and Anglophone audiences in different ways, negotiating potentially contentious concepts in his Chinese writings. In English he used the phrase *indigenous knowledge*, and this was a key part of the raison d'être of the NGO that he created in Kunming, the Center for Biodiversity and Indigenous Knowledge. For a time, CBIK was one of the busiest hubs at the confluence of the environmental winds and the blossoming interest in indigeneity in Yunnan, and likely in China itself. CBIK organized so many projects, both international and domestic, that they were said to be draining the regional pool of experts from institutes like the Academy of Social Sciences. Indeed officials contacted the NGO, insisting that they had to arrange more formal dealings with the research institutes from which they drew scholars. Unlike more generic phrases such as *traditional knowledge* or *local knowledge*, the phrase *indigenous knowledge* implies a specific connection to indigenous peoples and therefore has a potentially greater political edge, especially in places like China, where indigeneity has little purchase (Purcell 1998). When Xu wrote or spoke in Chinese, he used the term *chuangtong zhishi*, often translated as "traditional knowledge," which was less politically charged.[21] By using less politicized Chinese terms and operating within the rubric of scientific study, Xu's work was increasingly referenced by others, making it an important force in shaping new plans and approaches for conservation in Yunnan advocated by groups such as the Provincial Forestry Department and the Ford Foundation (Young 2005).

A Fundamentalist Communist: The Actions of Yu Xiaogang

Yu Xiaogang was part of the first wave of Yunnanese scholars explicitly interested in indigenous knowledge. He also became one of Yunnan's most confrontational proponents for social justice. In 1992 Yu received a fellowship from the Ford Foundation to study for his master's degree in Thailand, a country with an active history of indigenous activism. He soon returned to Yunnan to study indigenous knowledge for his thesis fieldwork. Most studies of indigenous knowledge seek indigenous groups living in places viewed as remote and at the margins of state control, who have developed a rich body of indigenous knowledge through centuries of place-based livelihoods. Yu, on the other hand, sought villages that were virtually at the other end of the spectrum: groups recently relocated by state authorities, who had removed them from the nature reserve.

Yu was concerned that the environmental winds had caused many hardships for rural peoples and that state authorities were following Western approaches that were unsuitable for China. In his report he frankly discussed the hardships now faced by relocated villagers, as their new village had poor-quality land and less access to clean water. The nature reserve staff, using terms familiar in contemporary China, had described these villagers as mere peasants or backward ethnic minorities. Yu strongly disagreed; he described them as sophisticated indigenous people. He was one of the first to add gender to his studies and documented "women's indigenous knowledge." He argued that these displaced villagers' indigenous status should ensure their rights to land tenure and greater decision making. He did not just declare them indigenous; he provided evidence of their identity and their special knowledge in detailed information about their use of medicinal plants. Yu's work resonated with the work of Xu and Pei, and he drew from their studies, showing how swidden can enhance biodiversity and how the Dai had created and maintained what Pei called "Holy Hills." His own style of engaging was quite different from that of Pei or Xu, however, who would not have chosen a site that was so politically contentious. Pei's and Xu's efforts were firmly situated within the realm of scientific research, and they avoided explicit social and political activism that could be construed as confrontational.

Yu's work was linked to and supported by a number of foreign organizations, such as ICIMOD (based in Nepal), the Ford Foundation, the MacArthur Foundation, and Oxfam–Hong Kong. His reports, in English and Chinese, circulated widely and were frequently cited as evidence that there was indigenous knowledge, and therefore indigenous people, in Yunnan (Yu 1993, 1994). This kind of transnational work was key to showing organizations within and outside of China that indigeneity existed—in large part by making it visible and evident through detailed studies, not just declaration.

Yu carried out fieldwork and worked directly with resource managers. At the time, many of these managers held to the dominant discourse about nature and society in Yunnan, which posited that ignorant, desperate, and scientifically illiterate peasants threatened the natural world. Yu's unprecedented advocacy for an indigenous space, however, made him a unique interlocutor. In one case, he carried out a survey at the Xishuangbanna Nature Reserve Bureau, the agency that had relocated the villagers. He asked the reserve staff a series of questions regarding indigenous knowledge, such as, "Why didn't they share more power and decision making with the indigenous people who fell within their purview?" Another part of the survey states, "The indigenous people have rich knowledge of [the] local ecosystems and environment; they even developed some strategies for sustainable use of natural resources." The interviewees were asked to register their opinion of this statement by choosing one of four answers: agree, disagree, neutral, or don't know. After several similar statements, Yu asked, and more or less promoted, "co-management" with indigenous peoples. This survey was unlikely to convert the reserve staff to following an indigenous rights orientation, yet his actions were aimed not just at discovering existing attitudes but at working to foster new ones. At the end of his report analyzing the results of his survey, he criticized the staff managers as having "little knowledge about indigenous agriculture practice" and said that some "even have a bias [against] indigenous culture (i.e., indigenous people are backward, primitive). [Staff] also have little knowledge about how to cooperate with [an] indigenous community in rural development" (Yu 1994: 23). He framed his work in nationalistic terms, saying that the Nature Reserve Bureau was adopting a "Yellowstone National Park model" that was unsuitable for China. Yu's argument, quite radical at the time, was that China needed to develop its own model of nature reserves that did not require the eviction of local peoples.[22]

In contrast to foreign expectations that Chinese activists would both confront the state and work surreptitiously, Yu aimed for visibility as he worked to invent new meanings and practices around indigeneity in China. He did not consider himself antistate but appealed to multiple levels of the state apparatus using vocabulary and tactics that were interwoven with references to Mao, Jiang Zemin, and minzu. Most surprisingly to many foreign observers, who imagine that Chinese activists fight to create a space for civil society away from the reach of the state (often understood as a place of freedom),[23] Yu both worked with and challenged government leaders by framing himself as more loyal to the unachieved aims of Mao than they. He often described himself as a "fundamentalist communist," a neologism possible only in today's China. The term punned on what he saw as a rising Christian fundamentalism in the United States and elsewhere and a corresponding decline of interest in communism in China, as the government increasingly turned toward the market economy in the 1980s. By the time I arrived in China in 1995, few others referred to themselves as communists. Mao-era terms that had been in daily use, such as *tongzhi* (comrade), were used mockingly; calling someone a good comrade signified that he or she had a naïve and misplaced sense of institutional duty; the term was also slang for "gay."

According to some of Yu's peers, he was among the most cosmopolitan of the people they knew, having traveled to many countries, and someone who knew how to speak to academics, bureaucrats, and high-level officials from around the world. Yet he was also of an older generation, who combined the hope that communism meant a better life for the masses with a sense of responsibility and an almost tireless zeal. In my conversations with him, his position as a fundamentalist communist seemed genuinely felt and also served to deflect criticism that he was too confrontational. But such a position had its limits.

Yu's efforts vacillated between direct confrontation and a more cautious approach. Although his English-language publications and reports used the phrase *indigenous people*, his initial Chinese texts focused on ethnic minorities (*shaoshu minzu*). By 2001 his Chinese writings were more likely to use the term for farmers (*nongmin*), which resonated more with older Marxist terminology and included the ethnic Han majority. In that year he published an article in Chinese in *Huaxiaren Wenhua Dili* (Chinese Cultural Geography), one of the new glossy journals aimed at entertaining an urban readership. The journal represented a striking contrast with the majority of state-backed publications, which, when they dealt with far-flung villages, tended to stress how the Party brought development and science to backward and poor groups. *Chinese Cultural Geography*, however, often describes pleasant and intriguing rural scenes, and the articles romanticize rural life, describing places where "colorful local culture survives." Yu's article was surprising because it did not fit this mold; it was more of a harangue than a celebration, an indictment of the difficulties faced by local residents living at a lake in Northwest Yunnan, especially after the lake was dammed, flooding their fields and curtailing their fishing. He had been conducting fieldwork there for his doctoral degree at Thailand's Asian Institute of Technology and at the same time was working for The Nature Conservancy.

After the article was published, he was called into the township government's office, where officials threatened him, saying that he had made them lose face. Yu, later recounting the meeting, told me that he quoted from Chinese leader Jiang Zemin's statement that the government must investigate and understand the peasant's situation, which slightly diffused the situation. He often quoted powerful government officials and encouraged villagers to do the same. This follows tactics used elsewhere in China, what Kevin O'Brien and Lianjiang Li (2006) call "rightful resistance," when groups challenge local governance by invoking national regulations. Around the same time, he organized residents living in the lake's vicinity to document their ecohistory and traditional knowledge, which included publishing articles in a local newspaper. These articles contended that the nature reserve staff's view was that local residents did not understand or appreciate the ecology and thus should face more restrictions. Yu hoped that the newspaper articles would enlarge the conversation by including the public rather than limiting it to a debate between the local residents and the reserve staff. He used such newspaper articles, field visits, and other means to build urban sympathy and support for rural livelihoods.

Yu represented a new form of subjectivity in China. He navigated a number of roles as a graduate student based at a foreign university, a staff member at a powerful international NGO, a founder of a small Chinese NGO, a social science expert, and a tireless social organizer and public critic who negotiated and challenged the boundaries of Chinese governance. While I knew him, his reputation as a social troublemaker in China was growing, but at the same time he was also widely respected and admired, especially among the more socially progressive expatriate and Chinese experts. In 2002 he rose to become China's most famous environmental activist, receiving a number of prestigious international awards. Later, however, he struggled to maintain his presence as an active voice, as his NGO came under increasing restrictions.

Yu's efforts helped me to question what it means to be an activist in contemporary China. This gap between how he was seen abroad (as a grassroots environmental activist) and within China can be partly explained by drawing on Naisargi Dave's (2011) insights about activism in India. As Dave argues, Western pundits often assume that the rise of gay rights in India follows a "flow model," that is, that the rise of gay rights in the West creates a general template, followed by "the rest." In other words, what counts as activism should follow narrow Western expectations, which ironically are not seen as Western but universal. Such pundits tend to see globalized formations besides gay rights as like a flow that starts in the West and spreads around the world. Dave also says that Westerners typically define activists as people who work for social change, often against forces of institutional power, such as corporations or the government. In other words, activists are usually regarded as those who practice organized resistance against powerful institutions and advocate social change considered politically progressive, such as feminism, environmentalism, or antiracism. This is also a narrow way to understand what activism is and can be.

In China I found that, like the term *indigenous*, the closest equivalent to the term *activist* has a radically different genealogy than in Western contexts. In China the most common terms for activists include *huodong zhe, huodong yuan,* and *jijifenzi*, which refer to those who actively promote state-mandated campaigns or state-approved projects. This is not how Yu saw his actions, for these are terms created by the state and used positively to describe people who carry out state directives. Yet, on the other hand, he did not fulfill the typical conceptions of activism outside of China, where it is generally assumed that activists oppose the state or at least challenge it to foster new kinds of social programs to ameliorate class, gender, or racial inequities.

In Chinese the best term to describe the people who directly confront state projects is one that was used during the Mao era and is still used to some extent today "counterrevolutionary" (*fan geming*). It is a serious term warranting arrest or at least punishment. Counterrevolutionary acts carry a charge akin to treason.[24] This term would never be used as a self-appellation, nor is it used by ordinary citizens; it is a threatening term wielded carefully by state officials who usually have significant stature.[25]

Although Yu's actions were sometimes risky, he strove to locate himself more firmly on the ground of rightful resistance, what in China is often referred to as being a *dingzi* (an iron nail). Socially the term *dingzi* is ambiguous and can be used in reproach, appreciation, or both. *Dingzi* describes those who challenge authority, including those who fight for their own personal welfare. Dingzi do not directly oppose the state but rather try to make their case by leapfrogging local authorities and gaining the attention of higher-level state officials. The group of social advocates most commonly described as dingzi are those who pursue HIV/AIDS rights, yet their efforts are not always successful (Shao 2006; Hyde 2007; Wan 2001). Some of these people have been jailed for their actions, especially those who revealed politically sensitive social disasters, such as how regional AIDS epidemics were created by entrepreneurs who carried out flawed blood-collection programs. Like the concept of indigeneity, the idea of rights for those who suffer from HIV/AIDS is politically contentious. And as with the term *indigenous*, the difficulty in finding simple equivalences across regional differences for terms such as *activist* should signal some methodological caution and a need to understand the social landscape in which such traveling concepts are being shaped.

Yu's earlier work, which focused on finding indigenous knowledge possessed by those evicted from nature reserves, started to expand as he considered the plight of those evicted by dams. He did not want to limit his involvement only to those groups that could be easily considered indigenous, those that more closely followed expectations of a "tribal slot." He took some of the strategies from an indigenous rights approach and used them in working with rural Chinese who were strongly affected by development projects in which, as is generally the case there, they had no input. His changing strategies shows us the continuing tensions over an indigenous space and the ways it might be morphed or stretched, or even abandoned by its previous proponents.

Yu's efforts led to some of China's first advocacy-based research on the social effects of dams, a topic already sensitive due to the often harsh criticism that Beijing had suffered, both domestically and internationally, over the Three Gorges Dam, the world's largest. Yu's peers cautioned him to suppress his results, which included the finding that one of the minority groups had turned to picking through garbage dumps after dam development took away their farming, hunting, and gathering lands were taken away. Instead Yu contacted reporters and widely publicized his results.

Next he brought villagers from potential dam relocation areas on "study tours" to visit villages that had already been relocated from dam sites. Government officials felt especially threatened by anyone trying to organize public campaigns of opposition and viewed such actions as dangerous (O'Brien and Li 2006). It is important to note that although China is now said to experience nearly 100,000 protests a year, almost all of these are described as spontaneous, not organized.[26] Although Yu's "study tours" encountered some difficulties, he was able to carry them out without being arrested, which was remarkable given the often intense state efforts to stop organizing beyond the local scale. However, in 2004 he crossed the line when he brought villagers from a proposed dam site to speak at an international meeting in Beijing. After bringing these dingzi villagers into an international site of expertise, his NGO was quickly shut down and his passport revoked.

Fortunately for Yu, this was not the end of his story. In 2006 he won the Goldman Prize, regarded by many environmentalists as the "environmental Nobel Prize" (Claudio 2007). Perhaps because of the award's international stature, the Chinese government gave Yu back his passport and allowed him to receive the award in California. Ironically, his status quickly changed from a national troublemaker to a source of patriotic pride, and Beijing sent a team from the state-run media to accompany him; it portrayed him as a Chinese hero of international fame. Three years later, Yu traveled to the Philippines to receive another major prize, the Ramon Magsaysay Award, sometimes called "Asia's Nobel." Even though this award bolstered his international standing, it also brought him greater scrutiny by officials, who were growing increasingly nervous about his international fame.

Yu's articulations with international networks and his growing prominence at major conferences provided him with a certain amount of support and leverage for his work, yet it also created new pressures. His work for rural people's rights, if not necessarily for an indigenous space, also became increasingly visible within China and furthered his reputation as a potential troublemaker. Ironically, others thought his previous calls for indigenous rights would alarm state officials, as these efforts contradicted a state policy that refused to recognize indigenous people in China. Instead it was his later work to improve rural livelihoods that drew more attention, even though such efforts were ostensibly part of state policy and a key part of the platform of PRC president Hu Jintao, who gained leadership in 2003. Yu's actions became even more confrontational when he expanded his mandate from working on indigenous issues to working with all villagers, whether or not they might be regarded as indigenous. In China organizing groups at levels beyond that of a village is an activity almost exclusively reserved for government officials and is often regarded by officials as a threatening act. In contrast, scholars who worked in a single village with a group potentially regarded as indigenous might be more easily ignored. Yu combined confrontational tactics from indigenous rights advocates with familiar techniques from Mao-era mobilizations, inventing new ways of being a dingzi and new understandings of rural livelihoods and rights in China.

In many ways Yu's potential for these new understandings and the emergence of an indigenous space in Yunnan were made possible by Pei's and Xu's earlier and ongoing work. At recent conferences, dozens of Pei's and Xu's colleagues articulated their passionate interest in sacred lands, indigenous knowledge, and the links between cultural and biological diversity. A decade earlier, Chinese scholars and project staff had overwhelmingly described the need to educate ignorant peasants and teach them scientific methods of farming, including how to use fertilizers and pesticides. This latter view of rural inhabitants remains dominant in Yunnan, but a number of socially progressive experts, like Yu, Pei, and Xu, are changing the social and political climate of conservation and development.

Their work in fostering an indigenous space is neither a top-down imposition of a foreign social category nor a spontaneous bottom-up activist social movement. Rather an indigenous space was being worked out mainly in an intermediate realm, pushed outward, tentatively and unevenly, by Chinese experts. They did not rally support for a roughly equivalent term to indigenous people, such as *tuzhu ren,* nor did they attempt to foster grassroots mobilization. Instead their actions point to particular interventions carried out in a Chinese context, creating new accounts of swidden agriculture, botanical reports, biodiversity statistics, and maps of sacred forests. For example, the switch from thinking of old-growth forests as evidence that ethnic minorities still stubbornly clung to feudal superstitions to understanding them as ecologically valuable sites in resilient sacred geographies has major implications for rethinking the relationships among peoples, landscapes, and indigeneity. Likewise the shift from slash and burn to swidden has enormous consequences for views on uplanders' agricultural practices. These efforts created the conditions of possibility for certain groups to be positioned as indigenous (Li 2000). The notion of an indigenous space challenges assumptions that certain groups' status as indigenous is self-evident and merely requires acknowledgment by governments or others. It emphasizes the creative work involved in what some call decolonization, which refers to both the tearing down of existing colonial structures as well as the simultaneous building up of new ones.

This approach also shows how this space emerges in relationship to shifting sets of social hierarchies. Indeed it would be difficult to overemphasize how much the language of indigenous knowledge and environmental sustainability in Pei's and Xu's reports deviated from existing social hierarchies. Many practices that were later seen as part of indigenous culture had been interpreted for decades as remnants of feudal superstition and were frequently subject to severe criticism and often attack. These experts' narratives and images worked to displace such earlier notions and provide new interpretations. Their work energized a new generation of scholars in China, who have now begun to search for and document Yunnan's "indigenous forest management systems" (Cheng et al. 2008; Liu 2006; Long and Zhou 2001; Liu et al. 2002). Many scholars and government workers now regard these systems as complex and culturally distinct institutions for sustainable natural resource use, worthy of study and, potentially, promotion.

SUMMARY

The making of an indigenous space in Yunnan is part and parcel of the environmental winds, and it is unfolding in diverse and complex ways. In part, this has been motivated by these experts' efforts to dislodge earlier conceptions of backward minzu and replace them with new forms of knowledge that promote an indigenous space. Their work has now affected the ways that other experts choose research topics, experimental designs, and comparative frameworks for projects on environmental conservation and rural development. Pei Shengji's early studies helped establish the concept of sacred lands in China, fostering the new concept that minzu have maintained biodiverse and sustainable forests and agricultural fields. Xu Jianchu's work documented the relatively benign ecological effects of swidden agriculture compared to dominant state-led land use, such as tea and rubber plantations. As he told me, he wanted to displace the image of "slash and burn as the bogeyman of tropical deforestation." Yu Xiaogang began by pushing for recognition of indigenous knowledge and ended up promoting greater social participation for all rural peoples, regardless of whether or not they were labeled or potentially labeled as indigenous. He thus used concepts from indigeneity as a provisional means of imagining alternative futures in which rural people's knowledge and rights were valued and respected.

Unlike much scholarship on indigeneity that focuses on established grassroots movements, looking at China reveals the difficult background work required for an indigenous space to gain traction in the first place. In this case, much transnational work around indigeneity in China was not seen as a separate issue of human rights or indigenous rights but as linking rural rights and environmental issues. This chapter reveals that there is not one indigenous movement that flows, extending a suite of global norms, but that diverging indigenous spaces are made possible by connections to and efforts by many involved in global indigenous movements.

Indigenous-identified groups are claiming a new negotiating position in relationship to global environmentalism, strongly disrupting patterns in how conservation has been carried out in the past (Brosius 2004). Such groups now show up at major environmental conferences and bring their complaints and demands to the bargaining table. These demands, situated within a larger field of advocacy and support, cannot be easily dismissed; they show how the conceptions and status of indigenous people vis-à-vis the environment and the state are not stable but are undergoing redefinition. In turn, governments, multilateral organizations, transnational corporations, and conservation organizations are facing new and heightened scrutiny when their work involves groups regarded as indigenous, because their organizations' actions are challenged by indigenous-identified peoples, indigenous advocates, or, as I have shown in this case, engaged experts. People trying to foster an indigenous space in divergent places around the world encounter quite different terrains.

Although much work on indigeneity supports the UN position that indigenous identities are based not on state definition but self-definition, in Yunnan most rural citizens—including those who would most likely be deemed indigenous by outsiders—have never heard of the concept. Self-definition of groups, not as particular minorities within state-recognized categories but as indigenous within globally recognized categories, is not always obvious or automatic but often requires extensive engagements with others.

Some people argue that the category of indigenous is foreign to China and therefore inapplicable. It is important to remember that less than a hundred years ago, the Marxist concepts of class and socialism were also foreign to China, but millions of Chinese now live in worlds deeply shaped by these concepts and practices. Socialism itself can be seen as a globalized formation that has strikingly divergent manifestations, from Cuba to Laos to the Soviet Union, and that continues to change. The concept of indigenous therefore should be understood likewise, as an ongoing process of conceptual exchange, circulation, and translation, providing new ways of understanding and acting in social landscapes. Some of these understandings become powerful and persuasive, mobilizing people like winds to work together toward alternative futures; other understandings never gain social force and wither away.

The concept of an indigenous space allows us to go beyond legalist approaches that rely on particular criteria (such as genetics, residence, or linguistic markers) to judge whether or not a certain group qualifies as indigenous. It refuses to naturalize indigeneity and instead encourages us to explore how different groups, in relationship to each other, invent, elaborate, and use this category. The concept encourages a curiosity in exploring the changing social lives of indigeneity, which have often transformed quite dramatically over the past decades. It does not assume that state recognition of indigenous status is an endpoint, or even desired by all groups potentially seen as indigenous.

Whereas much scholarship on indigeneity has revealed the struggles and the success of various indigenous movements, less attention has been paid to the work that makes the concept of the indigenous salient, especially in places with less public and state sympathy or places where, despite great efforts, it has failed to gain a foothold (Van Cott 2003; García 2005). In Yunnan the question of indigeneity emerged in connections with blowing environmental winds, winds that a number of people pulled in new and unexpected directions. And as we shall see in the following chapter, it wasn't only humans that were affecting and affected by the winds, but other species as well.

FIVE

On the Backs of Elephants

> We should think of wildlife as a relational achievement spun between people and animals, plants and soils, documents and devices in heterogeneous social networks which are performed in and through multiple places and fluid ecologies.
>
> WHATMORE AND THORNE,
> *"Wild(er)ness: Reconfiguring the Geographies of Wildlife"*

WHAT ROLE DO WILD ANIMALS PLAY in global environmental efforts? Previous chapters looked at how Yunnan's experts played a prominent role in shaping the environmental winds that blew through China. By engaging in various forms of selective engagement and transnational work, they refashioned their province as an important environmental hub within China and ultimately around the globe. Experts did not do all this work by themselves but enlisted the support of many actors, including villagers who maintained compelling environmental knowledge and were willing to describe it to outsiders. More surprisingly, experts' efforts were also boosted substantially by the presence of charismatic animals—specifically, the Asian elephant. Thousands of years ago these elephants lived as far north as present-day Beijing, but today they are found only in Banna. Elephants have survived millennia of threats from armies sent to capture them alive, commercial ivory hunters, and local vigilantes, as well as significant changes in climate and landscape.

This chapter explores how China's last remaining wild elephants—around two hundred in the year 2011—helped shape the environmental winds that were blowing through China beginning in the late 1980s. Both humans and nonhumans were swept up in these winds. One of their more prominent effects was a massive gun-confiscation campaign, aimed not at reducing crime but at eliminating hunting. Villagers told me that this campaign inspired the elephants to become particularly brazen and fierce. In one part of Banna, an area somewhat smaller than the state of New Jersey, approximately two hundred elephants killed at least sixty-three people in just five years (1997–2002), making them one of the most lethal groups of elephants in the world.[1] We don't know how many elephants survive, as the forest is thick, the elephants are elusive, and the territory is vast. China's forest elephants are rarely seen by plane and are not fitted with radio tracking or GPS devices. Since 1991 estimates of their population have fluctuated between two hundred and three hundred, based on rough counts, with few elephants seen. There are no reliable estimates of elephant populations before this time, but their habitat range was quite extensive more than a thousand years ago (Elvin 2004). Wildlife experts suggest that over a twenty-year period, these same elephants also devastated local farmers by destroying an estimated 100 million pounds of grain.[2]

In turn, a wide range of people, from villagers and government officials to animal welfare NGO workers, tried to accommodate the elephants in new and innovative ways. Elephants are not merely passive objects in environmental processes but are actively shaping and being shaped by the social and natural landscapes in which they live. It appears that the elephant population may again be growing in China, and while it remains difficult to enumerate populations in thick forest and rugged terrain, China's conservation success is being touted as a positive role model for other Asian countries. Situated within the revitalized academic interest in human-animal relations (e.g., Mullin 1999, 2002; Wolch and Emel 1998; Ritvo 1987; Haraway 2008), this chapter explores the role of elephants in transnational environmental work. I also use this chapter as a springboard to tentatively address broader notions of agency, both human and nonhuman.

APPROACHING THE QUESTION OF ELEPHANT AGENCY

In order for transnational connections to take place, human actors must carry out all sorts of tasks. People working for nature conservation do not perform only tasks solely taking place between humans (such as writing reports, conducting interviews, and trying to manage other humans' actions); they also try to mobilize particular plants and animals to do their work for them—as images or as part of narratives that create an effective field of attraction that will entice others to care about and support a particular project. Therefore I argue that it is not only humans who contribute to transnational work around globalized environmentalism; nonhumans also play significant roles, but not all play an equal role. Of the thousands of animal species in Yunnan, elephants may be the most important in terms of motivating transnational connections.[3] Especially since the 1980s, they are increasingly part of growing networks, connected with a number of organizations concerned about endangered species, elephants, Asian elephants, conservation, animal welfare, and human-animal conflict.

China's elephants, in turn, are enmeshed in globalized precedents.[4] Elephants have attracted international concern, and their treatment and trade, dead or alive, are now subject to scrutiny and regulation. In China elephants have become, like giant pandas, an important transnational animal. Elephants play a role in forms of transnational work that link scientists, government officials, farmers, conservationists, tourists, and others. I am not suggesting that elephants are *intentionally* carrying out transnational work toward a shared goal, but as objects of attraction, rallied by biologists and conservationists, they play a critical role in making "wild Yunnan" a place in the global ecumene. Elephants are used to elicit support from international conservation organizations, national governmental leaders, ecotourists, and others who build networks through them. Whereas in the 1970s China's elephants were mainly seen as a local problem and of little value, by the 1990s they were playing important and multiple roles as objects of touristic desire, threats to human livelihoods, and emblems of biodiversity. In these and other ways, elephants play a role in making global environmentalism what it is today.

Since the 1960s a number of scholars have written about the ways nonhumans are used as symbols (Rowland 1974; Douglas 1970; Leach 1964; Lévi-Strauss 1963). Yet nonhumans are not just semantic devices for human thinking. Likewise many people have written about animals as recipients of human action: passive research subjects for biologists, recipients of care by animal welfare groups, or targets of hunters who covet their tusks, pelts, or antlers. Animals are also more than recipients of such human actions; they also act themselves, and not only in response to human intentions. In contrast to many studies of human and nonhuman relationships that look at creatures confined to homes or zoos, this chapter looks at how wild elephants play an important role as agents in creating large-scale change.

Elephants are a physically powerful species that has gained the protection of a strong Chinese government. Elephants and humans respond to each other in a variety of social and physical ways, and coexisting is not always easy. Elephants wander, eat crops, collide into cars, and destroy houses. In

Yunnan they interact with people, crops, and other animals and plants, and people have changed their land-use practices and laws to accommodate elephants' actions. Elephants are increasingly becoming a source of tension for local people, who both admire their intelligence and will and fear their ability to destroy houses, fields, and human bodies. Villagers are frustrated because they have been stripped of the means to fight back (their guns), and elephants seem to be taking advantage of human helplessness. In these ways, elephants are playing a key role in coproducing new rural landscapes in southern Yunnan through what Jeremy Prestholdt (2007) refers to as "cumulative agency." I elaborate this concept below; for now suffice it to say that he understands agency as accumulated effect, not individual action.

Given all of the above, how might we understand elephants as transnational and biophysical agents, especially when notions of agency are typically predicated on questions of intent? How might we think about agency in elephants' engagements with farmers, researchers, and tourists, as well as with other animals and plants that occupy the same space? Agency is often defined as requiring conscious intentionality. Nonhumans are usually assumed to lack intentionality, so that attributions of agency to nonhumans are often dismissed as anthropomorphic. Rather than try to answer the question of animal intent head-on, I turn to ideas found in social and political thought that collectively help get at the question of animal agency.

One way that scholars are now trying to think differently about agency and relationships beyond the strictly human is through actor network theory, especially as pioneered by Bruno Latour (1987; Latour and Woolgar 1979). Latour attempts to dethrone humans' sovereign position by showing how authoritative knowledge is produced through an expansion of networks that link together a wide range of things—including humans, radioisotopes, sheep, and microscopes—all of them ostensibly equalized by the category of "actant": anything with the ability to act. The concept of the actant is helpful because it dislodges the human monopoly on agency, which is often seen as an intentional act of heroic individualism or collective will. It provides one way to write new accounts in which humans are not the only actors.

Latour asks us not to explore networks as given configurations, but to ask how they are made as historical products, to uncover how actants are successfully linked together, or "enrolled" in a network. To his credit, Latour stresses that actants may not behave as expected; they may not be obedient and docile subjects. For example, as famously described by Latour's colleague Michel Callon (1986), despite repeated attempts by French scientists and fishermen to replicate the conditions for the successful cultivation of scallops they observed in Japan, they could not domesticate them in French waters.[5] In the language of Callon and Latour, the scallops could not be enrolled in these new networks; they were unwilling participants.

Even though human efforts at enrolling other species, whether sheep, microbes, or scallops, are not always successful, humans remain at the center of these cases. It is difficult to imagine dislodging humans from the most active position in these accounts, although other objects now play active roles themselves. Although the concept is intended to show the range of actants participating in a network, in practice many scholars see humans doing all the work, building increasingly stable networks. Latour (1993) shows, for example, how Louis Pasteur worked with sheep and microbes to successfully create vaccines and came up with effective theories and practices to reduce the spread of infectious disease. All of us now live in Pasteurian worlds, with vaccines and pasteurized milk.

Another conceptual tool I use to help me approach the question of agency is Prestholdt's notion of "cumulative agency," which he discusses in his remarkable book *Domesticating the World: African Consumerism and the Genealogies of Globalization* (2007). By showing that trade was also deeply affected by Africans' desires, Prestholdt's account of colonial trade between Africa and Europe pushes us to rethink dominant notions of globalization as driven solely by Western interests. He argues that at a cumulative level, Africans' interest in *particular* objects (specific kinds of cloth, beads, guns, and so forth) powerfully shaped the specifics of the British economic and industrial landscape. In contrast, many accounts of global trade largely examine the social effects of trade on non-European locations but not Europe itself—the upshot being that globalization is seen as a one-way process in terms of social effect.[6] When Europe is seen as changing through these relations, this is often described in general and beneficial terms, such as Europe's increasing wealth or power, and it assumes that Europeans control the trade. Prestholdt, however, shows Africans' role in changing Europe, such as the way fortunes were made, where and when factories were constructed, and how the rhythms of labor were transformed. He moves away from typical conceptions of agency as the deliberate use of individual will toward specific aims and understands agency as dispersed but nonetheless socially powerful because of its cumulative force. My own work draws on a similar understanding of agency and appreciates his portrayal of global trade as both dynamic and, more unusually, reciprocal, which shows the making of globalization in ways that impact models cannot. The notion of cumulative agency is, like transnational work, a way of understanding change that is reciprocal and does not assume a fixed sensibility or goal but that can nonetheless have powerful consequences, as the combination of a thousand small, seemingly insignificant acts.

This chapter casts its net somewhat wider than Prestholdt does by trying to understand how humans and nonhumans participate in global formations, such as environmentalism, a movement that often justifies its existence, raises money, and recruits allies through the use of particular animals. Animals, including China's elephants, are now caught up in national, regional, and worldwide networks of care and concern. Elephants' everyday acts—eating people's rice and corn, destroying houses, threatening human bodies—should not be understood as resistance, but nonetheless their cumulative acts impact how humans organize themselves as well as how they literally shape landscapes and affect other species besides humans.[7]

We can also understand agency in ways that avoid the judgment of intent by focusing instead on effect, especially as a cumulative phenomenon. Rather than trying to determine if a particular animal action is or is not an expression of agency, we can start to see how nonhumans are lively in a way that social theory rarely recognizes (Whatmore 1999), though it is beginning to do so (Bennett 2004, 2009; Wolch et al. 2002).[8] Part of attending to this liveliness is seeing forms of social life as coproduced and understanding the ways that wild animals have memories and histories, not just instincts.

The activities of nonhumans shape the plans and actions made by humans to accommodate them. Understanding this can help us explore the ways that elephant lives are coproduced with other species, including humans.[9] The fact that elephants desire cultivated rice and corn more than wild bamboo in the forest is one critical way that their lives have been transformed in relationship to humans. In turn, the fact that elephants seek out farmers' fields has powerful effects on how farmers plant their fields, try to protect their crops, and interact with state officials. Regardless of elephants' intent, their continued presence in Yunnan is changing many people's lives, including how planners design and modify new roads, how nature reserve staff lobby to expand the size of reserves based on elephants' travels, and how scientists and elephant advocates carry out research and try to devise development projects to address human-elephant conflict. Thus a wide range of human lives and activities are influenced by their relationship to elephants. It is not only what elephants do directly as biophysical beings that matters, but the larger networks that are built around the elephants that effect change in the world on behalf of elephants.

Although animal behavior is usually assumed to be instinctual, automatic, and unchanging over time, one can challenge such assumptions by putting history back into the conception of natural history. It is often asserted that natural history and animal behavior are the result of a long-term evolutionary process, but as a neighbor in Xiao Long told me, the elephants, like his own peers, were adapting to China's social reforms, and doing so in particular ways. He described how elephant behaviors—how they find food, raise their young, and live social lives—are not fixed but are historically in flux. Elephants in the 1980s, he said, were different animals than those in the twenty-first century, when they became far more aggressive in fields and less fearful of humans.

The remainder of this chapter examines how China's elephants have been articulated within transnational networks, starting with the implementation of projects by WWF. I explore how elephants were part of changing social and natural landscapes in Yunnan, how they were different in 1980 and 2000, and how they were part of larger networks that were creating new worlds. I demonstrate the ways that elephants, through their liveliness, have been part of how transnational work is carried out during the environmental winds and how they are playing roles in transforming larger landscapes.

TRANSNATIONAL ELEPHANTS

There has been a massive rise in the number of transnational networks that have been formed around Yunnan's elephants. These creatures have been part of transnational trade and tribute relations in Asia for centuries, before nation-states were created (Wylie 2008). Before discussing how the arrival of the World Wildlife Fund in the 1980s meant new attempts to redesign human-elephant relations, I begin with a brief story about the first scientific scrutiny of these elephants in the 1950s.

Searching for Elephants: The PRC's First Biological Survey

When a group of biologists left Kunming in 1956 to travel to Banna to search for wild elephants, many of them suspected that they were too late. The journey to Banna had changed remarkably in a short period of time. The new PRC government had quickly pulled this frontier borderland—where, other than tea factories, there had been relatively little governmental control during the twentieth century—into national circuits. One of the scientists told me that in the 1940s, his trip from Kunming to Banna took twenty-eight days walking on a muddy track beside mules and horses. In 1956, however, he traveled by truck with other biologists on a new road for only a few days. He and the others were shocked and pleased to find that some wild elephants had survived (Shou et al. 1959). Their presence buttressed the biologists' successful petition in 1958 to create a nature reserve, one of the first in this new nation and the last to be created for nearly two decades afterward.

The scientists' pride was quickly challenged when, along with many others, they were punished and reeducated in 1957 during the Anti-Rightist Campaign. The Cultural Revolution was a time of massive political turmoil, with little official support for nature conservation. Much of the rain forest surrounding the designated reserve had been targeted for conversion into rubber and tea plantations, but the earlier designation as a reserve did provide some protection. According to Professor Yang Yuanchang, in numerous documents from the 1960s and 1970s there is no mention that the plantations might harm elephant habitat. Oral histories with local villagers revealed few details about elephant encounters during this time, and almost no written materials on elephants discuss this era. It seems that many rural men were armed, and they would shoot at any elephant that threatened their fields, with little fear of retribution from authorities. There were no forest guards or regulations against hunting, and there were still bounties for wild tigers, which were regarded as a pest to be wiped out.

By the 1980s, however, interest in nature conservation began to rise again, spreading beyond a small group of older biologists to become part of a larger social force, part of the environmental winds that were blowing through the region. Around that time a new frame, "the environment," began to connect existing and emerging concerns about wildlife, air and water pollution, deforestation, sand storms, erosion, floods and water scarcity, and so forth. These winds led to a number of new efforts to conceive of and address ecological problems in the grasslands, desert, and forests. In Yunnan, with its intact forest and diverse plant and animal life, these winds fostered the creation of a new wave of nature reserves, the first in decades, and the resurrection of earlier reserves, such as the one created to protect the elephants and Banna's tropical rain forest.

WWF Engages with Xishuangbanna's Elephants

It was the existence of this nature reserve in Xishuangbanna that had attracted the attention of staff from the WWF. If all of the tropical rain forest had been converted to rubber, tea, and other crops, it is unlikely they would have shown any interest. It should also be remembered that the elephants' survival, which represented substantial efforts on the elephants' own part over the long term, also helped foster the creation of this reserve in the first place. In 1986 it was the surviving elephants that had motivated Prince Philip and WWF staff to justify applying for and carrying out a series of projects (Hathaway 2010b). In turn, WWF's involvement helped stimulate the proliferation of transnational connections with China's elephants, integrating them with already extensive and active networks that included other Asian elephants along with their scientists, conservationists, and advocates.

WWF endeavored to reconfigure human-elephant relationships with respect to two distinct groups: rural villagers and urban tourists. On the one hand, it supported the state's campaigns against villagers' traditional slash-and-burn practices, viewed as damaging elephant habitat, and it aimed to separate villagers from elephants by creating permanent fields using agroforestry. WWF also assisted villagers in their efforts to keep elephants off their farms by providing them with electric fences. On the other hand, WWF tried to lure urban residents to the nature reserve to watch wild elephants as a form of tourism. To this end, they built China's first nature tourism center in the early 1990s, which was premised on tourists' attraction to elephants. WWF staff saw this as a way of creating a new dynamic: a place for watching elephants in the wild rather than in cages or performing in shows. In so doing, they hoped to cultivate new sensibilities toward wild animals more generally and to foster in the Chinese a sense of love for nature that WWF staff from England, the United States, and Hong Kong all believed was sorely missing.

Building Fences: Separating Elephants from Rural Chinese

As described in chapter 3, after WWF's agroforestry plot was severely damaged by domestic water buffalo, staff brought in barbed wire for villagers to build a new fence. Although the fence worked as intended, villagers pointed out that it was relatively worthless against animals such as wild pigs and elephants. At this point, in 1991, WWF had worked for several years in Banna and had gradually become aware that elephants were a major threat to village fields, including their experimental agroforestry plots. With much fanfare, WWF imported electric fences. They brought in miles of wire, a number of solar chargers, and batteries for several sites. WWF had much experience with electric fences, a technology designed to separate elephants from humans' desired spaces, having already used it in African projects for some time (Thouless and Sakwa 1995).

WWF's use of electric fences was a form of transnational work, bringing experiences in African countries to a new social context in China and a different species of elephant. Africa was not only the center for agroforestry but had also become a major center for the innovation of techniques to divide megafauna from rural villages, in part because Africa was one of the earliest sites for wildlife conservation, dating back nearly a century. Crop predation by elephants continues to be a critical challenge for elephant conservation, as angry villagers throughout the world fight back using deadly methods such as poisons or the use of lethal high-voltage power lines.

During my fieldwork in 2001, villagers in Xiao Long told me about their initial response when setting up the electric fences around their rice fields: "The Americans [WWF] said it was very expensive; it was Western technology, Western science. It seemed very good." The first night, after the elephants were surprised and backed away from the fence, the villagers were quite pleased with their luck. The elephants returned the next night, with the same effect. Within a week, however, one elephant had pushed a tree down on the wires, thereby short-circuiting the whole fence. After tentatively feeling around the wires with their trunks, the elephants walked over the fallen barrier and into the croplands, where they feasted on rice. Each day villagers reinforced poles, repaired the wires, and removed trees that were close to the fence line. This went on for weeks, with the elephants returning every few days. Eventually, however, one elephant took a stout stick in its trunk and swung it like a bat at the fence. With this weapon, the elephants once again succeeded in destroying the fence, and did so repeatedly. It seemed as though there was no way to counter the elephants, and soon the fence was abandoned.

I was surprised to hear that the fences were abandoned so quickly. Archival materials in WWF's Beijing office and publications by the Xishuangbanna Forestry Bureau (Tang et al. 1998) mentioned plans to use electric fences but did not describe the results. These fences were first put up in 1991, and still no satisfactory replacement had been found by 2000. Living in Xiao Long, where the presence of elephants was often quite terrifying, I started to do my own research on alternative methods for keeping elephants out of fields, mainly by reading about efforts in Sri Lanka and Africa. One night, while drinking rice alcohol with five men who had just regaled me with tales of elephants, I told them that in my cursory study, I had found that electric fences were actually fairly effective in Africa. I wondered why they did not work here. One of the more humorous men rose from his chair and started to walk toward the door, pausing to look over the fields that he and others had tried to fence a decade before. "Well," he said, "what can I say? These are *Chinese* elephants. Maybe they are smarter than those African ones." Some of the other men sitting around a wooden table looked at him and then, smiling, nodded at me.

During fieldwork I was struck on a number of similar occasions to learn that Xiao Long residents had seemingly contradictory feelings about elephants: there was a sense of fate and helplessness toward elephants eating their crops, and at the same time there was a sense of pride and awe about the elephants' abilities. While conservationists write scientific papers that represent elephants as static entities with fixed habitat needs and instinctual behaviors, local people understand them as an agentive, strategic, intelligent, and creative species. Villagers see elephants somewhat like themselves, in that their lives are structured more by desires than by a notion of mechanical needs. Furthermore elephants are said to not merely react to human practices but to experiment themselves; they are not only victims of human actions but instigators themselves.

Making Wild Elephant Valley

At the same time as WWF was trying to keep rural citizens away from elephants, they were also planning to bring urban Chinese and foreign tourists to the elephants. To do so, they reached out across the Pacific to hire tourism experts from New Zealand. In 1988 WWF staff contacted Tourism Resource Consultant's Dave Bamford, who assessed the potential of what was then called "nature tourism." (This was before the concept and popularity of ecotourism had emerged.) He was the first specialist hired for the project. In the fall of that year, Bamford (1988: 3) spent a month in Yunnan, where he felt he was starting practically from scratch: "To date there has been virtually no tourist use of the Xishuangbanna Reserves despite the fact that the main road from Simao to Jinghong passes through the middle of the Mengyang Reserve. Near Mengyang there is an artificial elephant pond built by the Forestry Bureau but neither elephants nor tourists use it. The Reserve personnel have had little experience with tourism and are understandably apprehensive about the potential impacts of tourist developments."

Although Bamford thought the reserve personnel were apprehensive about tourism, he also imagined that they might appreciate it as a new source of funding. Federal support for the reserve was sparse and unpredictable. Bamford hoped that the elephants would be an effective draw, and he saw foreigners as the primary audience, even though WWF was hoping to effect change mainly among Chinese urbanites. He noted that in just nine years, from 1978 to 1987, the number of foreigners visiting Kunming had increased a hundredfold (from 1,024 to over 130,000). He was not able to obtain statistics on how many tourists went from Kunming to Xishuangbanna, but it was estimated that 3 to 5 percent of them came south and that domestic tourists probably came in twice these numbers. This implied that for 1987, there might be as many as 6,500 foreigners and 13,000 Chinese tourists coming to Banna. He also suggested that current tourism conditions were far from adequate and far below international standards. He stated that getting from Kunming to Xishuangbanna was quite difficult: "The 700 km. trip to Jinghong is notorious for its rough roads, poor accommodation and inadequate visitor facilities. Air transport to Simao, the nearest airport to Xishuangbanna, is infrequent and unreliable. The plane is an old (>40 year-old) 40 seater Dakota.[10] The 4–5 hour bus trip from Simao to Jinghong is through spectacular hilly and lowland valleys. Many people become travel sick however" (Bamford 1988: 21). Nonetheless Bamford described his plans for creating an elephant viewing area: "At a popular animal watering-and-feeding area one hour's walk from San Cha He (a nearby village), a simple tree house will be constructed. Animals will be encouraged into this area by spreading salt on the river bank" (21).[11] Bamford's plans were carried out and then quickly expanded. By 1991 workers had built several buildings and a parking lot only a few minutes from the wild elephant viewing area. Yet despite the salt, which the elephants greatly desired, and the expansion of a bathing area, elephants showed up only seasonally and sporadically. At that point the park had relatively little to offer them, as the surrounding landscape had hundreds of places for elephants to bathe and a number of natural salt licks. Also at that time humans were well-armed and most elephants were still fairly nervous about being close to them. As elephants were not easily enrolled in WWF's plans, many tourists were disappointed, and the site was threatened with closure.

WWF decided to take action. Although the original intention of the ecotourist center was for visitors to see wild elephants, this wasn't working out. WWF staff then attempted to acquire tamed elephants. As China lacked a tradition of elephants and mahouts (trained elephant riders), WWF turned to Thailand and Laos. Luckily for WWF, Thailand had recently become a country with a massive surplus of unemployed elephants. Thousands had previously worked in upland logging camps but lost work following a government-imposed logging ban in 1989 after a devastating flood. Although floods had happened before, this was the first time new environmental logics blamed upland deforestation for lowland flooding and resulted in a ban. Even though the ban resulted in many unwanted Thai elephants, bringing them to China was not easy because they were still considered an endangered species. Trucking them across borders required special permissions from organizations that had been created in 1973 by the Convention on International Trade in Endangered Species (CITES) and that enforce the international trade of endangered species. The Laotian government quickly approved the arrangements, but for Thailand, WWF had to resort to "high-level diplomacy" using staff from WWF-International's headquarters in Geneva (Santiapillai et al. 1991: 13). Ironically, WWF, which had played a strong role in strengthening CITES regulations, now found these laws hampering its own efforts. Nonetheless WWF was successful. They did not want the tourist site in Yunnan to fail, even if it meant importing trained Thai elephants and providing visitors with elephant shows.

China's Elephants Become Asian

At the same time that WWF was trying to import trained elephants, it was also sponsoring research on wild elephants, thereby linking them with global conservation networks for the first time. Before the early 1990s China's elephants had been more or less left out of these networks, and the elephants were still largely known as Indian elephants because India had the largest elephant populations and most active research community. As Asian networks expanded, the elephants were renamed Asian elephants. There were two prominent Asian elephant conservation groups with many delegates from India and Southeast Asia, but as late as the early 1990s none came from China. Douglas Chadwick's celebrated book of 1994, *The Fate of the Elephant,* contained only one sentence about China. Previously WWF had relied almost entirely on existing Chinese ecological data and hired relatively few foreigners to conduct research. However, as there was little recent information available in Chinese or English on China's elephants,[12] WWF hired an internationally prominent elephant researcher, Charles Santiapillai, in 1991.

Santiapillai was working for WWF-Asia's main office in Bogor, Indonesia, after carrying out research in his native Sri Lanka. As he remarked, Sri Lankans displayed a remarkable tolerance for elephants' continued presence, despite a long history of human-elephant conflict and ongoing civil war. Before he traveled to Xishuangbanna, he had worked in Sumatra, where he had experimented with the tremendously difficult task of relocating elephant herds. In 1990 he and Peter Jackson completed a survey of Asian elephants that was the first to provide a country-by-country analysis and conservation action plan.[13] At that time he interviewed Professor Yang Yuanchang, the same man who had accompanied Prince Philip. Yang suggested that there were likely only 150 elephants left, many of them "migrants from Laos" (Santiapillai and Jackson 1990: 23).

Santiapillai and Jackson's survey was an important first step in linking China's elephants with international conservation efforts. Santiapillai discussed China in the 1991 meetings of two of the most important networks for elephant conservation in Asia, the Asian Elephant Specialist Group of the International Union for the Conservation of Nature and the Species Survival Network, which predated CITES. China had not yet sent delegates to these groups, which mainly consisted of researchers from India and Southeast Asia, but Santiapillai's studies and participation at these meetings, however, helped to connect China's elephants, scientists, and policymakers with these emerging transnational networks that were centered in Asia.

Santiapillai's fieldwork relied heavily on testimony from Chinese experts and villagers. His method therefore contrasted starkly with those of previous foreign experts in China, such as George Schaller, the most famous giant panda biologist in the Anglophone world. Schaller spent long periods of time alone, studying animal populations in remote places that were least affected by humans. Santiapillai worked with a Chinese scientist from the Southwest Forestry College (Zhu Xiang), who acted as his English translator, and two staff from the Xishuangbanna Nature Reserve (Hua Dongyong and Zhong Shengqin). The team did not attempt to find elephants deep in the reserve's center but traveled to sites where villagers reported elephant encounters. They mapped out sixty-four locations, estimating the population at around five hundred. They discovered that no elephants had been observed west of the Mekong River and that several areas once thought to contain elephants probably did not.[14]

Santiapillai viewed elephant research in China as far behind work done in Sri Lanka and India. He proposed to bring his three Yunnanese colleagues on a study trip to Sri Lanka to learn about census measures, and to southern India to study elephant management. Such exchanges were emblematic of the movements of people and techniques that were key to creating a trans-Asian conservation network. The incorporation of Chinese elephants into globalized environmentalism was not a self-propelled flow from a Western center. Instead the transnational work of Santiapillai and others focused mainly on Asian circuits and involved the movements of people and funds, equipment, and even live animals that were not always willing and obedient subjects.

Unlike WWF's main project, which had denounced slash and burn as the main threat to elephant habitat, Santiapillai argued that rubber plantations were a more important problem. These plantations meant that forests were permanently converted into monocultures, eliminating food and cover for elephants. He also noted that raw rubber was processed using firewood and suspected that this was a major source of deforestation. Santiapillai took an unusual position for a conservation biologist at the time: he saw swidden agriculture as relatively benign environmentally and viewed farmers less as aggressors toward and more as victims of elephants. As he and his coauthors stated, "Resolving the conflicts between man and elephant is one of the principal concerns of the Forestry Bureau in Xishuangbanna" (1991: 8).

WWF's Project Evaluation and the Environmental Winds

It turned out that Santiapillai's positions anticipated some larger shifts. The environmental winds that encouraged strict nature conservation efforts were influenced by a rising interest in rural social justice and new ecological perspectives, which emphasized the positive role of "disturbances" caused by human or natural events, such as storms. These shifts were quite visible in 1995, when the European Union's evaluation team declared that WWF had completely misunderstood the agricultural situation: villagers in Banna were not backward farmers practicing slash and burn but people carrying out indigenous shifting cultivation. According to the evaluators, WWF had been wrong to advocate agroforestry and introduce potentially invasive, exotic species. Instead, the EU said, local people should teach WWF about sustainable livelihoods using native species (see Chapter 1). In response to the EU evaluation, WWF crafted a new proposal in 1996 and hired Chinese experts to diagnose previous problems and create new approaches.

The new proposal contained a dramatic reversal in several aspects, especially in terms of the relation between swidden agriculture and elephants. The old project promoted agroforestry to keep people on the farms and out of the forest, with the idea that slash-and-burn agriculture harms forests and elephant habitat. Following this perspective, WWF had tacitly supported efforts by the Nature Reserve Bureau to relocate over a thousand villagers from the core zone. In the new proposal, villagers were to be paid to go into the reserve, including the core zone, and carry out shifting cultivation with the explicit aim of increasing the food supply for elephants. Thus the new plan saw shifting cultivation as a positive force, an activity that enriched elephant habitat through the provision of food not only directly from rice and corn but also indirectly through the growth of the tender wild plants that appear after plots are no longer cultivated.

I think it is fair to say that this radically different proposal was in part stimulated by the elephants themselves. WWF and government officials realized that human-elephant conflicts were increasing, and they hoped that hiring villagers to commute to the reserve's core and grow food for elephants could decrease conflicts in the outer villages. Elephants' actions had confounded WWF's previous efforts and, in turn, shaped conservation plans in China, much as they have for decades in Africa and elsewhere. Similarly WWF staff's disappointment that their electric fences (thought to be effective in Africa) had been defeated by China's elephants directly shaped ongoing forms of transnational work they engaged in: collaborating with officials and villagers to negotiate elephant conservation that resulted neither in villagers' protests against a government that seemed to care less for them than for elephants nor in villagers' vigilante justice against elephants.

Although WWF's proposal for a new conservation project indexed a wind that was starting to blow around the world, connecting conservation with issues of rural social justice, it was not part of a paradigm shift among all of the world's institutions. As is the fate of most project proposals, WWF's new project did not gain funding. In China and many other places the stigma of slash and burn largely remained, and the new proposal did not indicate a wholesale shift in attitudes toward this practice. Even without the new WWF project in place, China's elephants were becoming increasingly enmeshed in transnational relations.

Based on my critical readings of international development programs, I anticipated that as a powerful Western institution, WWF would have had a substantial impact. I gradually learned that while WWF's effects continue to reverberate, most people agree that elephants have a far larger effect on farmers' everyday lives and landscape practices. In Xiao Long one day, I listened to a heated discussion about the elephants and WWF's project. Some of the younger generation referred to it by the term "project" (*xiangmu*), while others referred to the people who staffed it, calling them "foreigners" (*waiguoren*) or "Americans" (*Meiguoren*). An older man said, "Elephants are much more important than the Americans. The Americans came and went, but the elephants are here. We don't have guns, we can't defend ourselves." For a while he had believed that WWF, with its access to millions of dollars, the best science, and many workers, could keep the elephants at bay, but ultimately it could not.

This is not to say, however, that WWF was entirely inconsequential. Its involvement indexed new ways for villagers and officials to understand and manage the environment, which affected daily life more and more. After WWF left Banna in 1996, elephants played an increasingly important role as the key symbol of the state's growing emphasis on environmental protection, as well as becoming a contentious object of state responsibility to its citizens after elephant violence rose unexpectedly. Indeed it was not just that the elephants were powerful in and of themselves, but in part it was their link to state power that made them such a significant force.

How Elephants Are Changing Yunnan's Landscapes

During the 1990s and into the new century, elephants played a significant role in changing social and natural landscapes in southern Yunnan, far beyond WWF's project sites. These landscapes were reshaped by three main groups: state-based institutions, villagers, and the elephants themselves. The forces of conservation mandates were quite uneven; places like Banna, which had thick forests, nature reserves, and state-protected animal species, were sites of major transformation, whereas neighboring regions to the north that were seen as agricultural zones were only slightly affected. On the other hand, when elephants started traveling beyond Banna's northern edge, they inspired another series of projects aimed at reducing conflict between rural villagers and elephants.

If one were to focus on the role of state institutions in changing the everyday politics of nature, one might describe the following: Nature Reserve Bureau staff and the police began to increasingly scrutinize and criminalize everyday activities by villagers, such as grazing their livestock, collecting fuel wood, and farming by shifting cultivation. Hunting was banned, and villagers were given posters with information on which wild animals were now granted state protection, with elephants being the most valued. Police caught and executed four elephant hunters in 1995 and later confiscated many thousands of guns from rural villagers.

Although all of these events were quite important in transforming this political landscape, there were other actants involved, including elephants and rural villagers. Next I explore how actions by the elephants themselves, and actions by others on behalf of elephants, were negotiated in the daily lives of rural farmers. The changing landscape of laws, including the ban on guns, shaped the ways that villagers interacted not only with elephants but with other animals as well.

Increasing Chinese Interest in Elephants

Villagers in Xiao Long said that at first it seemed that foreigners, such as the American WWF staff, were the only ones interested enough in elephants to be willing to travel down the village's long muddy road. Yet by the mid-1990s they started to hear about a rising interest in elephants among Chinese, evidenced in the growing success of Wild Elephant Valley, which, in contrast to the WWF consultant's original plans, attracted relatively few foreigners. The site did, however, become well-known as a place for *laoban* (bosses) from Jinghong, Banna's main urban center, and was popular with tourists from Beijing and Shanghai. The biggest event in Xiao Long was the arrival of a Beijing-based TV film crew asking about elephants. The crew came unannounced, and when they saw an older man chopping firewood, they immediately started to film him. Recalling this years later, Li Baiwen, became upset. The old man, her father, was bare-chested and barefoot. They hadn't given him time to dress, as they should have before taking a photo, so they had "stolen" his photo (*touxiang,* photographed him without his permission).

Like other Beijing reporters in Xishuangbanna, the crew perceived it as a place of deep forests and ethnic minorities, an exotic landscape where people either lived in harmony with nature or threatened it.[15] The reporter asked

Li's father if elephants lived nearby or were a nuisance, and then asked if he shot at them. According to Li, the crew seemed intent on portraying her father as a mountain hick, and they kept asking him to repeat himself, making faces as if they were frustrated trying to decipher his difficult local dialect. She thought they seemed to want to portray their village as people "out of place" and locked in conflict with elephants, as this was an animal that the state and now urbanites wanted to protect. The crew's presence and questions, said Li, were part of a larger change, wherein the government and city people began caring more about the welfare of elephants than people like herself, the ordinary people (*laobaixing*).

Executing the Elephant Poachers

As further evidence for this shift of showing more concern for elephants than rural people, Li Baiwen told me that the government was starting to execute farmers accused of poaching elephants. The execution of four men was a landmark legal case in China, which occurred in 1995, while I was living in Kunming. Passing the provincial courthouse one day, I saw posters of the accused. Their heads were bowed and shoulders sunk in the classic stance of China's convicted criminals. In another photo, their guns were laid out, evidence of their crime. These capital crimes were later recorded by Amnesty International, which attempts to monitor state executions around the world; this was the first case in which elephant hunters made China's list and, as far as I know, the first time in China when killing an animal resulted in execution.[16]

I later learned that these elephant hunters had passed by Xiao Long. One day a forest-savvy older man, Zhang Li, who had grown up in another area and been taught much about hunting and trapping, took my family into the forest to see where the hunters had passed. As we climbed the trail, he stopped periodically to blow a buffalo horn. My son, at age three, despite exposure to American children's books and their portrayal of elephants as mild and friendly, had already gathered that his "uncles" and "aunts" in Xiao Long were fearful of the animals. He was afraid that the horn was meant to attract the wild elephants. "No," replied Zhang, "it is not to call the elephants—it is to let them know we are coming, so that we don't surprise them." He likely did this to add drama to our outing, as I never saw anyone else use such a noisemaker.

Zhang told us that five years earlier he was near this spot when he saw a gang of men carrying rifles. He knew instantly that they were hunting for elephants, for some of them had semiautomatic military rifles, which are expensive, hard to acquire, and illegal and usually possessed only by urban-based hunters who shoot illicitly for markets. He was nervous when he saw them, and watched from afar. This was the same gang later convicted of poaching elephants.[17] Some said that the gang tried to work with villagers, acting almost in vigilante fashion by targeting "rogue male" elephants that were particularly brazen about entering village paddy fields.[18] Soon after they were seen walking above Xiao Long, the gang was arrested in a police raid and taken to court. Especially because the gang used village trails, the executions left a particularly strong impression with Xiao Long residents. Even years later villagers were eager to recall the events, which demonstrated that the state had a new level of seriousness in protecting its elephants. Even so, villagers were still surprised that people had been executed, given that the death penalty seemed fitting only for crimes against fellow humans, not animals. Even though they described elephants as one of the "state's animals"—several species that, along with wild oxen, are now privileged for protection—this harsh action was unprecedented. As one old man said about the executions, "This would have never happened under Chairman Mao."

Guarding against Elephants

Humans have long had to deal with elephants, but recently the dynamics of these engagements have changed rapidly. During the 1960s cadres began the first phase of the "from the uplands to the lowlands" campaign, trying to convince small upland villagers to move into larger centers closer to newly built lowland roads. Villagers were also organized away from family-based farming into communal-style collectivized farming, and shared tasks included guarding crops from wild animals, especially elephants. Grain crops took about three months to grow, but the last few weeks were the most challenging, when the ripening grain was said to create a powerful fragrance that blew through the surrounding forest. Those who guarded fields, especially the ones that were far from the village, earned communal work points, which could be exchanged for grain and petty cash. The guards were almost always men, in part because this was seen as a lonely, difficult, and dangerous job and one that favored gun owners. Guards were supposed to stay awake all night on a bamboo platform with a small fire built on a mud base and use split bamboo torches to investigate incursions.

During the late 1970s the introduction of a new technology—flashlights—changed the relationships between guards, fields, and animals. Previously the guards had used light from a fire or torch, and used guns mainly to scare off wild animals rather than kill them. Flashlights were expensive and unreliable, using batteries that were hard to obtain and quickly ran out of power, but when they worked, a skilled hunter was better able to shoot animals, such as wild pigs or deer (*jizi*) that came at night. Unlike light from a bamboo torch, the beam from a flashlight would entrance some animals, making them easier to shoot (called "jacklighting" in English). Thus flashlights and guns helped transform fields of ripe grain from a liability into an opportunity to obtain meat, as nocturnal animals were attracted to the grain. The flashlight, however, did not significantly change relations with elephants. At first, elephants were apprehensive about a flashlight shining in their eyes, but unless it was accompanied by a bullet, they soon became unperturbed.

During the 1980s the communes were dismantled, reducing the number of swidden fields in the forest that were collectively managed. Fields tended to become smaller and more dispersed, and the burdens of guarding fell to the responsibility of individual families. It was therefore less likely that every plot was guarded each night. Villagers still worked together to guard the main paddy fields in the village center, where members of several families maintained a vigil. When the elephants appeared, neighbors were woken up to bang pots and pans, shine torches and flashlights, and light fireworks. Eventually yelling and banging metal became ineffective, as the elephants largely ignored the noise. Bamboo torches, which had previously made elephants nervous, now seemed to incite their curiosity, and elephants were said to be attracted to them. People tried different kinds of fireworks, but the favorite, which always makes elephants retreat, was a stick of dynamite. While dynamite was common for a while, given out by state agents for help in building roads and clearing boulders from paddy fields, by the 1990s it was becoming scarce. Dynamite started to be made in some small factories, which created less reliable and more dangerous sticks. In the house we slept in, the grandfather kept four old sticks of dynamite on his window ledge—kept "just in case," he said—their wrappers fading in the sunlight. One problem was that the lit wicks could go out halfway, leaving a precariously short fuse for a second attempt, and many were nervous about using it against elephants. In a neighboring village, one young man was using dynamite to stun fish in a river pool; after relighting a failed fuse, it exploded and he lost his hand. Much safer and more effective than dynamite was the gun. While elephants became immune to most loud noises and regular fireworks, all of them were greatly bothered by bullets, although bullets seldom killed them.

Yet by the 1990s shooting at elephants had become dangerous, as they became seen as the animals with the greatest degree of state protection. Some villagers described these nighttime events as a "battle over rice," wherein humans were fighting elephants in rice fields. Elephants were threatening in and of themselves, but were even more threatening because they were linked to the state in powerful ways. The elephant became an adversary that one could scare but dare not kill, for fear of being sent to jail or worse.

Elephant Trunks and Contraband Guns

By the early 1980s the environmental winds prompted a series of laws aimed at curtailing hunting and gun use in Yunnan, and these laws were strongly motivated by the desire to protect elephants.[19] Reports for the MacArthur Foundation, however, stated that despite the laws, hunting (almost always for animals other than elephants) remained widespread (Ma et al. 1994). Even though the sale of new factory-made guns and factory-made bullets was fairly well-regulated, few rifles were purchased in stores. Most rural citizens had old muzzleloaders. Guns were also fashioned by inventive craftsmen in small workshops that circumvented attempts to control their sale. State officials began to limit access to critical supplies, such as lead and gunpowder. Throughout the 1980s most bullets were made by hunters at home. They purchased slabs of lead at state stores, scored the slab with a knife, melted these chunks over a hot fire, and poured the liquid lead into molds to make shot.

As the government clamped down on bullet supplies, lead became a black market commodity and increasingly expensive. Hunters turned to other sources, discovering an excellent substitute in ball bearings. Mechanics now found themselves in possession of a commodity that, until recently, lay scattered about the shop floor or in drawers. The price of used ball bearings shot up overnight. Gunpowder too became increasingly scarce in some places. Hunters attempted to fashion gunpowder with new kinds of charcoal and sources of saltpeter. It was not an easy task, and hunters experimented with various combinations of improvised gunpowder and ball bearings of different sizes. This often made their guns less powerful and predictable.

Starting in the late 1990s the police carried out a massive gun-confiscation program throughout Banna that resulted in much hardship for the rural people, significantly changing their relationship to elephants. It was part of a larger campaign throughout China, especially concentrated along the western frontier (Harris 2008). It was not aimed at the reduction of urban crime and the acquisition of handguns; rather it aimed to strip rural people of their hunting rifles. According to older men and women in Banna, guns had been widespread for over a century and long played an important role in social life. Especially before the founding of the PRC, guns were needed for defending their possessions from bandits (*tufei*).[20] Many men carried guns in their daily travels to the fields, strapped to their backs or their bike racks, hoping to shoot wild game, such as deer or wild boar. Groups of men would also go hunting together at night or wait in the forest under large fig trees, where the ripe fruit attracted game. Wild game meat was highly valued, and the division of the meat fostered sociality between families. As mentioned, guns were also the primary form of crop protection against animals, particularly elephants.

In 2000 I was visiting a village when the police drove their jeep onto the basketball court, spoke to the leader, and made a public announcement. They called for all guns, and they even produced a list of people and the guns they owned.[21] The police's knowledge of specific guns was quite disturbing to local residents, for it indicated that their neighbors had informed the police about who owned what guns. The police stayed for a few hours as the villagers brought their guns to them. When the police departed, with the back of the jeep piled with guns, a few older men were clearly very disturbed, watching their prized guns carried off, some of which had been in their family for three or more generations.

That night we watched the news. One segment showed the police taking guns from Jinuo Mountain, where some Xiao Long residents had relatives. Jinuo Mountain had a reputation, even among foreigners, for challenging state authority and being hostile to tourists (Liou et al. 2000: 737). Someone commented, "Look at how nervous the police are. They're scared of those Jinuo people. They wouldn't dare go into their homes." The next scene showed police putting guns in a pile, drenching them with gasoline, and lighting them on fire. "The pile looks small—I wonder what will happen to my gun?" said one man. There was suspicion that many of the guns would not be destroyed but would instead eventually be sold by the police on the black market.

I later read in the local newspaper that police claimed to have confiscated over sixty-two thousand guns in one campaign (Anonymous 2001).

It remains to be seen whether the confiscation campaign will put a serious dent in elephant hunting, which to my knowledge was quite rare. Although obtaining homemade guns was relatively easy and widespread, these guns were largely ineffective against elephants. Some men reported that large musket balls would simply bounce off the elephant's thick hide and skull, especially balls propelled by homemade gunpowder. I heard stories of old elephants that were still strong after being shot by dozens of bullets.[22] Unlike wild pigs or deer, elephants were not killed for food, even though some heard rumors that wealthy cadres in the city feasted on braised elephant trunk.

In the 1970s dead elephants were of only local interest, and killing an elephant in order to protect family and fields was not a major crime. With the advent of the environmental winds, however, elephants became increasingly articulated with a growing assemblage of people and organizations, their life and death a concern of the forest police, the Nature Reserve Bureau, scientists and researchers, and a number of domestic and international organizations. If an elephant was killed, its body was so massive as to defy easy disposal. It was hard to keep a dead elephant hidden for long, as the corpse would soon attract vultures and other carrion eaters. When an elephant died, humans could smell its corpse from a long distance away after a few days in the hot sun. I was told that when dead elephants are found these days, the police investigate their cause of death just as when someone important dies under suspicious circumstances.

Elephant Agency and Trunks as Weapons

Elephants were not merely passive symbols or actants in these articulations but agentive in a number of ways, including how they moved through the land, ate cultivated crops and wild plants, interacted with humans and dogs, and crossed highways and international borders. Ironically, as a growing number of agencies became involved in their protection, the elephants seemed to assert themselves more frequently in violent ways. During my fieldwork, conflict was a prominent theme in stories about elephants and humans; it was easy to spend hours drinking home-brewed rice alcohol until late in the night, listening to men swap stories about their encounters with particular aggressive elephants.

In one account, said to have taken place in the mid-1990s, a couple was driving home on a small forest road with a walking tractor (*tuolaji*) pulling a trailer, where the wife sat after a long day of hoeing their crops. They came across a large male elephant with a broken tusk, standing off the road a bit. They attempted to drive past, but the elephant charged. The woman ran into the forest, while the man dove under the tuolaji. The woman watched as the seemingly enraged elephant lifted up the tractor, seeking her husband, who quickly rolled and crawled out of its way. Finally the elephant reached under the tractor with its trunk, pulled the man out, and stepped on him, crushing his chest and killing him instantly. The elephant slowly wandered away, and the woman, still trembling hours later, cut through the woods to her village, staying off the trail. She rarely left her home afterward.

Although some elephant advocates claim that elephants will never use their trunk as a weapon, local residents regularly used the verb *kao*, meaning "to hit, beat, or torture," to describe how an elephant's use of its trunk. They said that when elephants attack humans, hitting with trunks is the most common method, followed by trampling as a close second.

Women were less likely to swap their own tales of elephant aggression, but many could recount stories about others killed or wounded by elephants, such as the one described earlier. Stories were heard on the TV news. Some villagers, such as Li Baiwen, were wary of the media, suspecting that the news reported only a few elephant attacks, perhaps because the state was such a strong elephant advocate and officials were worried that reporting every attack would increase people's collective animosity and fear.[23] They did, however, learn about one of their allies, a reporter, who went to investigate where an elephant had raided a field. The reporter was killed after that same elephant leaped over a trench designed to protect the field and trampled the man.

There was a sense that over the past decade, elephants were getting more brazen. As well, elephants seemed to act out of spite: several neighbors reported that the elephants destroyed their *wopo* (also known as *woheng*), a temporary structure for eating and sleeping while tending their forest fields. Other crops, such as the medicinal plant sha ren, which the elephants did not eat, seemed nonetheless to be deliberately crushed over a large area. People said that the elephants knew that they were no longer armed; hence the elephants had become braver and more aggressive toward humans working or walking in the forest. People became more nervous about walking through the forest to visit relatives or attend their forest fields. Women openly admitted that they felt less safe in the woods, fearing elephants more than snakes or swarms of giant underground wasps. The children were also afraid of encountering the elephants and were more careful to walk to school together in a group, hoping that the elephants would hear them coming, and if they were charged, they could scatter in many directions. There were numerous animals to fear; we saw cobras and lethal giant walking stick bugs. Yet, as one child pointed out, such poisonous creatures could only ever hurt one of them, but an angry elephant could kill them all. Most villagers might go months at a time without seeing any elephants, as they stayed hidden in the forest, though their tracks were evident alongside the road and trails. During the fall, however, when the grain was ripening, the elephants were much more likely to come to the nearby fields in pursuit of food.

Elephants and Agriculture

The presence of elephants has transformed agriculture in a variety of ways. For example, WWF consultants promoted agroforestry and encouraged Xiao Long residents to grow corn under their planted trees. Many residents were reluctant, however, fearing that elephants would destroy the tree saplings in their attempt to eat the corn. They said that mixing valuable tree seedlings with crops was too risky. Due to what people perceived as increased aggression on the part of the elephants, people began to feel more nervous about tending their remote swidden fields or patch of sha ren. Some decided to give up plots that were particularly far away or necessitated travel in places where rogue males were rumored to frequent. Others avoided traveling at dawn and dusk, when the elephants were most active, meaning that people were spending less time taking care of their fields, leaving later in the morning and returning home earlier at night (Secretariat 2004).

It remains to be seen how the elephant presence will continue to shape evolving uses of land and family incomes, but in 2004 one report submitted to UNESCO's Man and the Biosphere (MAB) program by Chinese investigators suggests that as wildlife protection becomes increasingly effective, villagers are becoming poorer. The report stated that "local people think the local government only cares about wild elephants and not local people" (Secretariat 2004: 8), which echoed what I had heard from Li Baiwen, the woman whose bare-chested father was interviewed by the Beijing reporters, and from some of her friends.

Clever Killers: International Funding and Animal Rights

During fieldwork I was at times overwhelmed by people's sense of inevitable conflict; they were increasingly vulnerable to elephants that were becoming more numerous and gaining in courage. This conflict was exacerbated, some thought, by the ways that outsiders, from WWF staff to state officials, had shifted their concern from villagers' welfare to the welfare of wild animals. One of the most direct consequences of this shift was the confiscation of villagers' guns. On the other hand, because the elephant now held a special status, it meant that nature reserve staff were responsible for elephants' damage and were required to compensate farmers for their loss. The same was not true for other wild animals; when wild pigs or insects ate crops, there was no way to get reimbursed. The net effect of the compensation program (designed mainly for elephant damage) was to tie rural people more closely to the Nature Reserve Bureau so that its power over its "kingdom" was demonstrated and reinforced. Bureau staff decided which cases were and were not deserving of attention. Compensation was often a long, drawn-out affair, in which farmers had little appeal if the staff challenged their claim. Overall, farmers had little to gain, as the rates of compensation were lower than the cost of buying replacement grain.[24]

As much as the villagers resented the elephants, they were also amused and impressed by them. I expected that villagers would express more fear and anger toward a beast that threatened to kill or maim them or could destroy their entire crop in a single night. Yet I heard many stories about the ways that elephants were "naughty" (*tiaopi*) and "clever" (*congming*), the same terms used to describe well-loved and intelligent children.

I was surprised to hear some of these terms used by Shou Bin, an older man well known for his hard luck. Shou brought me to his field to take pictures, which he would later bring to the Nature Reserve Bureau in the hope of getting some compensation. As I was the only person in Xiao Long with a camera, photographing damage became one of my duties, and this was the second time Shou had asked for my help. As we approached Shou's field above Xiao Long, we could see his corn stalks knocked down in the mud, which was riddled with deep elephant tracks. I knew his family had little money, and the compensation rate was low and the payment often delayed.

He picked up a corn cob, exclaiming, "Look at this, He Wei. The husk is completely peeled back. Without hands, how can an elephant do this? They are very clever."

Yang Bilun and several others said that the best way for me to watch clever elephants in action was to visit Wild Elephant Valley, WWF's ecotourism site. A number of residents had visited there over the years, which was surprising to me, as many people had little cash and rarely traveled anywhere for sightseeing. But it had become a favorite destination for special occasions, such as an outing for newlyweds. The site was a favored place for posing with an elephant for a photo, which was then laminated and brought home as a souvenir.

In 2000 I accompanied my Xiao Long host family to Wild Elephant Valley, which was completely different from Bamford's original plans in 1988 for a simple tree house. The site was now located on tourist maps and was part of mainstream tourist packages. Tour buses regularly brought loads of tourists from Beijing and Shanghai, who traveled on day trips from Jinghong, the main city in Xishuangbanna. Wild Elephant Valley included a number of buildings, an aerial tram, a staff dormitory, several tree houses for guests, a butterfly exhibit, a large aviary, and an elephant show. It was no longer under the management of the Nature Reserve Bureau but subcontracted to a number of Chinese companies.[25]

Like many visitors, my host family was most interested in the elephant show, which was still performed by Thai elephants and trainers. Villagers had told me that the performance was delightful. The elephants played basketball, danced, and had enough money sense to know the difference between 10 and 20 yuan bills. The show seemed to prove that these beasts—whom they feared in the forest and hated and admired in the fields—were indeed very clever.

For the Nature Reserve Bureau, the elephants are a key way to attract international funding, as well as a daily headache and a continual liability. When I returned to China in the first decade of the twenty-first century, I found that elephants were increasingly entangled with new international donors, as well as receiving much greater domestic funding. Leaders at the Xishuangbanna Nature Reserve described how garnering international funding builds their prestige and helps them to gain attention from the central government, such as attaining the highest level of nature reserve designation. Thus, even though international support is less predictable and more bureaucratically challenging than domestic funding, it nonetheless offers important

FIGURE 14. City bosses (*laoban*) pose with tamed elephants in Wild Elephant Valley, 2001. Xishuangbanna, Yunnan. Photograph by Michael Hathaway.

symbolic capital. The Bureau has also piggybacked on international projects, using these opportunities to build in more training sessions and hire staff with greater English-language ability, who can then better negotiate and engage with donors. The constant pursuit of projects is essential to the reserve's continuing operation.

In some cases, international projects mandate publicly accessible reports, which provide insights that might not be available elsewhere. One report by China's Secretariat to UNESCO's MAB program highlights the increasing antagonism with elephants. It provides evidence of far more cases in which elephants have killed humans than can be gleaned by newspaper articles, supporting Li Baiwen's suspicions that not every death is made public. Over a five-year period, from 1997 to 2002, elephants killed at least sixty-three people, a disturbingly high figure for a population of elephants estimated to be as few as two hundred (Secretariat 2004: 8). As far as I can tell, this means that China's elephants have among the highest per capita "murder rate" in the world.[26]

Members of the Nature Reserve Bureau have been attempting to ameliorate these tensions in a number of ways. They have prioritized assistance for villages that have suffered from elephants, mobilizing various forms of aid, such as fee waivers for their students and construction of protective ditches and more electric fences to surround fields. They have requested and received more money for compensation from the federal government. They also worked together with local government officials to devise an insurance contract for elephant damage to fields and people, which started in 2009. While similar to previous compensation schemes, it goes beyond local-state relations to bring in private companies to further mediate these interspecies relationships. In 2010 villagers in Banna received nearly one million yuan from the insurance company. According to one article, the program has now expanded to include "all endangered wild animals under state protection—including elephants, boars, bears, tigers, and leopards" (Xinhua 2011).[27] These dynamics are changing the ways that insurance, liability, and agency itself is understood. As insurance companies are a relatively recent phenomenon in China, liability questions are an active site of growth and debate.

Even though elephants have precipitated the further development of compensation schemes, both public and public-private, there is a sense that this route will not be the final answer. There is increasing interest in using barriers to separate villages from elephants. The MAB secretary stated that electric fences work successfully in India, Sri Lanka, and Vietnam but that these fences have not yet worked in Banna because of "improper use and management." In other words, these officials believed the problem was in human error, unlike Xiao Long residents' belief that their "Chinese elephants" will always find ways to overcome the electric fence.

The report takes a "new ecology" view of elephant habitat, which departs from the older perspective that all human action in the forest is inevitably detrimental. Previously many reserve staff believed, like WWF, that suppressing forest fires and slash-and-burn agriculture would protect rain forests and elephant populations. The report, however, stated that thirty years of suppressing fires and slash and burn (now described as "nomadic agriculture") has actually damaged elephant habitat. In the 1980s the only positive voices for such forms of agriculture in Yunnan were a few Chinese experts, such as Yin Shaoting and Xu Jianchu (whose work was discussed in the previous chapter). For some time, their arguments were attacked or ignored, but by the late 1990s they had gained strength, and this report shows that they not only convinced many researchers and officials but motivated new approaches to environmental projects.

The report states that "nomadic agriculture" helped create important "food gardens" for elephants. Thus, as with WWF's 1996 project proposal to pay villagers to plant crops for elephants, shifting cultivation was reimagined as a way to improve elephant habitat. Indeed even though WWF was unable to obtain international support for its project, the Nature Reserve Bureau is now doing exactly that, with money from Beijing, creating what state officials call "dinner halls" for elephants (Anonymous 2006). I also heard that others were using creative methods to bolster China's elephant population, such as placing salt at the Laos border to entice elephants over the line into China.

Not only is the Nature Reserve Bureau hiring villagers to carry out swidden agriculture for the elephants, but elephants are the focus of massive plans by the Ministry of Forestry, amounting to an expenditure of over 130 million yuan (approximately U.S.$20 million). The plan includes creating corridors, approximately two kilometers wide, to allow elephants to travel between isolated nature reserves. A survey revealed that some reserves didn't contain elephants, so the corridors could enable elephants to repopulate there, and conservationists are trying to figure out ways to guide elephants into using these routes. Lately a massive new nature reserve connecting China and Laos was created for elephants, so these animals are inspiring new forms of transnational negotiation.

Inspired by China's Panda Center in Sichuan, which has become a mecca for biologists and tourists, there is also much state support for creating an Asian elephant breeding center, which would be part of Wild Elephant Valley and financed, in part, by WWF. If built, the center will create a cluster of veterinary expertise, a place for testing new diets and disease treatments, as well as a place to raise elephant orphans.[28] Growing support for animal rights in China may now shape the way that elephant shows occur in Wild Elephant Valley; for example, a 2010 law mandates improved conditions for captive elephants, with the threat of banning shows in facilities that don't meet new requirements (Moore 2011). Although it remains to be seen how this regulation will be enforced, laws like this represent the first legislation in China to improve the rights of captive or domestic animals. Chinese elephants, despite their small population, have motivated a wide range of transformations throughout China, traveling far beyond their own relatively small foothold in southern Yunnan and affecting many ways in which people and nonhuman animals interact and engage. One can see the kinds of cumulative agency that this group of elephants produces, as well as the kinds of interactions that they inspire.

SUMMARY

The story of human-elephant conflict describes a far more complex set of relations than is usually assumed. There is not just an adversarial conflict between humans and elephants but a complex relationship that includes a whole range of actants, including organisms such as rice, corn, dogs, and oxen and material objects such as guns, electric fences, and dynamite. These interactions have changed over time. We can see how the elephants' actions work cumulatively to have important effects on human lives as well as physical and social landscapes. Such transformations are happening directly and are also mediated by assemblages that extend beyond national boundaries being built around elephants. China's elephants are now increasingly enmeshed in networks, some of which, first formed by the "repentant butchers" of Africa in the early twentieth century, are now thick with conservation biologists, animal welfare organizations, veterinarians, members of zoos, and major circuses (like Barnum and Bailey), and they have now expanded to include Asian elephants.

In terms of the global elephant population, China's elephants are quite marginal, in part because their numbers are so few (around two hundred out of a potential total of fifty thousand Asian elephants in thirteen countries). As well, Asian elephant tusks are relatively small, so they play a fairly insignificant role in the thriving global ivory trade (although Chinese are becoming key players, as both sellers and buyers, in the African ivory trade). Nevertheless the legacy of transnational elephant networks is important in shaping the roles of China's elephants, and this is working out reciprocally. Thus we should not assume that China is merely the recipient of funds and projects from elsewhere, that globalized environmentalist or animal rights sensibilities are now flowing into China. At times I am reminded of the 1960s and 1970s, when China was an international model for social reform, anti-imperialism, and revolution. Since the first decade of the twenty-first century, Chinese wildlife managers have been successfully touting China as a global model for elephant conservation, so its own strategies for managing these relationships are starting to spread into other countries.[29] As I have shown, however, the stories of China's elephants are far more complex and challenging than clear-cut narratives of conservation success. The elephants and the people who live with them are rapidly changing through their engagements with each other and through the environmental winds, which are shaping these dynamics as well. People and elephants are the two most powerful landscape-changing animals in southern China; they are social, clever, and willful and part of moral and legal hierarchies. Elephants are neither domesticated animals nor wild beasts; their lives are far more intertwined with ours than is often imagined, and they are becoming different in the process.

Conclusion

AFTER I RETURNED to Ann Arbor, Michigan, from my dissertation fieldwork, I heard from a Chinese friend that her mother had just arrived. My friend didn't have a car, so I offered to take her mother, Li Ming, shopping so that she could buy some items that weren't at their nearby Safeway. Many of my friends in urban China were of Li's generation, newly retired professionals, and I quickly felt at ease with her.

We drove through town en route to an Asian foods market. She saw one grocery store and asked me if we should try there. As I translated the name of the store, the People's Food Co-op, into Chinese (Renmin Shiwu Heszoshe), I could feel my own sense of surprise, but her expression was even stronger: "What? In America? I thought America was capitalist. You *still* have people's cooperatives? We got rid of all our cooperatives a long time ago." I wanted to explain the co-op's place in U.S. history, but before I could catch my breath and talk about the American 1960s and 1970s, she had already moved on to the next topic, as we had already broached several strange American customs.

Her comment stuck with me for years. I had shopped in co-ops for years and had always assumed their legacy was from the American co-op movement that was strong in during the 1920s and 1930s. I never thought China might have had anything to do with the People's Food Co-op, even though its name was as common in Chinese as the People's Liberation Army or the People's Republic of China. Yet as I've slowly discovered over the years, in bits and pieces, the creation of these food co-ops was often intimately influenced by stories about China. Stories, books, and speeches about China and even visits there influenced all kinds of social formations. The feminist movement, the civil rights movement, and the Black Power movement were all deeply influenced by revolutionary winds from China and elsewhere. The language of revolution, the language of liberation, patriarchy, consciousness-raising, and creating a different world, the challenge to authority, the critique of capitalism and imperialism, and the reference to America's "Cultural Revolution" were just some of the connections with China. I, like many others, had basically understood the 1960s as happening in a national space, contained within national boundaries. There might be some coincidental simultaneity, such as May 1968 in Paris and Berkeley, but I had to search hard to find discussions of these transnational connections, the ways ideas, imagination, and people traveled across oceans, over mountain ranges, and across borders that made the global 1960s and beyond. They did not just travel by themselves but needed human labor or transnational work to translate them literally and metaphorically, to make them viable, and as they were reshaped in new contexts they became part of new traveling winds.

I later talked to some of the staff about the early days of the co-op, asking if China was on their minds when they started thinking about creating it in the late 1960s. One man told me, "Oh, certainly, China was so big. Everyone we knew was talking about our revolution, quoting Mao—like the hippies, the Black Panthers, the Marxists. Some of the old Marxists I met were happy to see what China was doing then; they were done with glorifying the Soviet Union. They saw the USSR as a case where socialist utopia failed and saw China as the next hope." These winds of change, created by the active efforts of many people who challenged the war in Vietnam, sexism, racism, and imperialism, were already blowing throughout the world before I was born, making me and the society I grew up in.

It is interesting to think that the existence of food co-ops in the United States is just part of the landscape now, despite the transnational winds that contributed to their initiation, and most do not question how they got there. Perhaps sometime in the not too distant future the "indigenous" category will seem equally natural in China, and the complicated, transnational, and environmentalist paths through which it became a viable category will be largely forgotten.

We are all shaped by the winds we live through, whether the Great Depression, the 1960s, or their aftermath. This is a different way of seeing the world compared to typical understandings of forces like modernization, globalization, and neoliberalism, which are often imagined as forces fully formed in the West, flowing outward and spreading throughout the world.

Scholars have long been thinking about the impact of globalization on places like China, just as they have long talked about the impact of the West and capitalism. As China becomes one of the global superpowers, however, some people have started to question the assumption that China is only on the receiving end. A few even invoke Mao's statement that the "East Wind will conquer the West Wind," which is no longer about socialism and capitalism but about China and the West. In the early twenty-first century it is now possible to see that the world is being reshaped by a Chinese-centered flow. However, I would like to make two points clear. First, I am not trying to make the argument that instead of Westernization, we are undergoing a period of Easternization or even a China-based transformation. Rather I have emphasized that winds do not blow in a straight line and often push up against one another, which makes them unpredictable in their force and direction. As people engage with these winds, both the winds themselves and their diverse sources are changed in an ongoing dynamic. Second, it should also be clear that I do not see these winds or globalization as new or unprecedented, for China's engagement with the world has been influential for some time, with its role in global trade, alongside the spread of ideas and technologies such as gunpowder and the compass.

With the notion of winds, however, we can start to investigate how these globalized formations travel and work in relation to the terrain they are blowing through. We don't know how environmentalism and indigenous politics will work out in China, but we can see how they are being shaped by the existing landscapes and are creating new ones. Winds always suggest a perspective of transformation, not fixity; of multiplicity, not singularity, that does not start and end within national boundaries or always begin in the West, but is made and remade through a thousand ongoing engagements.

NOTES

INTRODUCTION

1. Following standard Chinese orthography, I write Chinese names as surname followed by given name. I refer to people by their surnames. In romanizing Chinese names, I use the pinyin method, which was created in China during the 1950s. Chinese who published before this time used an earlier method, called Wade-Giles, to write their names. I add this in parentheses. I use pseudonyms for everyone except those public figures whose work is so distinctive that attempts to disguise their identity would be fruitless, or those who expressly wished to be identified.

2. In this book I mainly discuss natural scientists, but I also write about some social scientists. When discussing natural scientists, I use the term "scientists" (*kexuejia*), but when describing groups of natural and social scientists or those with training in both, I use "experts" (*zhuanjia*), a common term in China. Zhuanjia designates those with expert training, usually an undergraduate or advanced degree, who possess some degree of authority. Although in anthropology there has recently been a strong critical stance against "experts" and their claims to authority, I am sympathetic to this particular group, a number of whom are, like myself, also concerned about social justice and do not always exhibit the kind of hubris and arrogance that many tend to associate with experts.

3. The term "wind" has a long history in China, and is still frequently in use. I draw on several aspects of this legacy, while also departing from conventional understandings of feng in several ways, as I explain in the subsequent chapter. The historian Shigehisa Kuriyama (1994) suggests that, in ancient China, "[t]he imagination of winds reached beyond medicine and meteorology to encompass ideas of space and time, poetry and politics, geography and self" (23). He explains how winds "foreshadow change, cause change, exemplify change, are change"—which is similar to the way that I use the concept (24). Other scholars, such as Hu Houxuan (1944), Chris Low, and Elizabeth Hsu (2008) have examined the diverse understanding of winds around the world, but surprisingly, there is little scholarship on the concept of winds in modern China. There are many expressions containing the

term wind in recent Chinese history, such as great wind, Right deviationist wind, black wind, spring wind, east wind, west wind, cold wind, warm wind, evil wind, new wind, communist wind, exaggeration wind, levelling and transferring wind, and darkest wind, as well as many wind-related metaphors to describe rapid social change such as a whirlwind or hurricane.

4. In China, the most common term is "foreigners" (*laowai,* literally "old outsiders," or *waiguoren,* meaning "outside nation people"), whereas "expatriate" is a more common self-appellation for outsiders who live in China for a long period of time.

5. There are a wide range of accounts on China's environment. Smil's book title, of course, is an ironic inversion of Pearl Buck's famous novel *The Good Earth* (1931), which helped Buck win a Nobel Prize for literature in 1938. Some of Smil's arguments that claim the destruction of China's natural world is a direct result of Mao's policies were paralleled in Boxer (1981), and echoed in Judith Shapiro's *Mao's War against Nature* (2001). In addition to Smil, a number of other influential accounts of China's environment posit different causes and times, but many argue that China became a "bad earth," an environmental wasteland. Historians often go much further back in time, and scholars such as Peter Perdue (1987), Mark Elvin (2004), and Robert Marks (2012) explore China's long durée, arguing that severe environmental damage had been occurring for centuries.

Others, such as William Hinton and Elizabeth Economy, focus on environmental tragedy after the beginning of the reform era (starting in 1978). Hinton, a long-time supporter of Maoist projects, sees the reform era as precipitating "a wholesale attack on an already much abused and enervated environment, on mountain slopes, on trees, on water resources, on grasslands, on fishing grounds, on wildlife, on minerals underground, on anything that could be cut down, plowed up, pumped over, dug out, shot dead or carried away" (1990: 21). Economy, in contrast, while not looking at the Mao era with rose colored glasses, shows a wide range of post-Mao environmental problems in her book *The River Runs Black* (2004). I find value in each of these accounts but am skeptical of overly generalizing narratives and claims about the condition of or causality behind China's environment writ large.

My analysis mainly draws on a number of exciting and important new studies on China's environment and environmentalism since the beginning of the twenty-first century. These include monographs by Chris Coggins (2002), Richard Edmonds (2012), Richard Harris (2008), Anna Lora-Wainwright (2013), Bryan Tilt (2009), Robert Weller (2006), Dee Mack Williams (2002), and Xie Lie (2012), as well as important articles by Peter Ho (2001, 2003), Setsuko Matsuzawa (2011, 2012), Ralph Litzinger (2004, 2006), Wu Fengshi (2009, 2012), Yang Guobin (2005, 2007), and Emily Yeh (2005, 2009).

6. The term "blue ants" was popularized starting in the 1950s and 1960s; see Paloczi-Horvath 1962, Guillain 1957.

7. Steven Flusty's *De-Coca-Colonization: Making the Globe from the Inside Out* (2004) is a fascinating account, one of the first I read arguing that global connections are made through social action. Flusty, however, sees "global cities" such as

Tokyo and Los Angeles as the main nodes for these globalized connections, whereas I explore how such connections are made far outside the city. He provides a more general synopsis, whereas I focus on the transformation of environmentalism as a specific globalizing dynamic.

CHAPTER ONE

1. There is, however, increasing concern about China's water and air pollution from sources such as factories, coal-powered electrical plants, and cars.

2. Such formations come into being, spread, travel, and transform not only through changing discourses or social movements but also through legal interventions, the creation of formal and informal institutions, and diffuse, everyday social and spatial reconfigurations, including transformations in how people behave with each other. For example, we can look at the U.S. legislation Title IX, passed in 1972. This law was originally part of civil rights' efforts to dismantle Jim Crow racial discrimination in hiring and retaining employees. Title IX used the mechanism of withholding federal financial support from institutions, especially schools, that had discriminatory policies. One of its primary unforeseen effects was to expose tremendous inequities in boys' and girls' sports programs in public schools and mandate reform. In turn, this led to significant changes in millions of children's social worlds, building thousands of gyms, creating thousands of girls' sports teams, and changing orientations toward the body, exercise, diet, and gender relationships writ large (Bolin and Granskog 2003). Such transformations were not just created de novo or intentionally but came about as part of often overlapping movements for civil rights and women's rights that emerged in the 1960s, in part through globalized politics.

3. I also heard other metaphors to describe the rapid rise of environmental concern in China such as a "green hurricane" (*luse jufeng*), the same feng meaning wind. One example of this use comes from Qu Geping, the head of China's first Environmental Protection Agency, in the documentary *Waking the Green Tiger: A Green Movement Rises in China* by Vancouver-based director Gary Marcuse (Marcuse 2011).

4. Indeed the Cultural Revolution is often described as a whirlwind, storm, or hurricane.

5. Although the Great Leap is often described, not totally inaccurately, as a major disaster (see Hershatter 2011; Han 2008), it should also be recognized that the vast irrigation systems built during this time increased grain production and therefore the food supply and wealth for millions of Chinese; many of these systems are still used today.

6. It should be noted that by 1986, the starting point of this book, environmental laws and interests were already part of Chinese worlds. The first federal environmental laws were established in 1979, and in 1981 some universities created environmental law departments.

7. Historically China's eastern region, with powerful cities such as Beijing, Nanjing, and Shanghai, had a stronger legacy of governmental institutions that regulated natural resources such as forestry and fisheries (Songster 2001; Muscolino 2009). Many of these institutions started after the fall of the Qing Dynasty in 1911; for the next forty years, China faced civil war, invasion by the Japanese, and negotiations with foreign powers, which effectively reduced its capacity to govern. In western China, state powers were even more limited, with less reach.

In places such as colonial Africa and India, a long history of European policies regulated Africans' and Indians' legal access to wild animals, trees, and land. Many animal species, especially large ones such as elephants and tigers, were claimed by the state and completely protected or actively managed, with networks of game wardens and guards, hunting licenses, and other regulations (Anderson and Grove 1989). By the beginning of the twentieth century, laws for game animals (i.e., those sought by sport hunters) were in effect throughout much of the world.

8. There are ancient conservation legacies in China, such as imperial game parks and temple forests, but it is difficult to determine how these were actually managed (Menzies 1994). These older examples are often quite different in orientation from explicitly conservationist programs.

9. For years I tended to believe the account of this campaign as ecologically devastating and that killing sparrows created massive insect outbreaks. More recently, however, I learned that far fewer birds died; rather than estimates of many millions, some authorities now say that 800,000 birds were destroyed by "hysterical crowds." To put this in perspective, however, such carnage pales compared to the birds killed in the United States by domestic and feral cats. MacKinnnon and Philips (2000) estimated that these cats killed several million birds a day, as many as one billion in a year, and a paper in *Nature Communications* claims that free-ranging domestic cats kill 1.4–3.7 billion birds and 6.9–20.7 billion mammals annually (Loss et al. 2013).

10. In one of the many fascinating, little-known stories about China's global connections during the Mao era, a time almost always characterized as one of international isolation, wild animals played an important role in generating foreign connections. The most well-known example is "panda diplomacy," whereby China used gifts of live pandas in fostering improved relations with a number of countries. Also, however, China conducted a brisk trade in animals' parts such as crane feathers, pelts from wild cats (including tigers), and musk glands from wild deer, coveted in making French perfume (P. Li 2007). Sales of wild products, especially to Japan and Europe, were especially critical as China struggled under a U.S.-imposed trade embargo.

11. This last cutting particularly alarmed officials, who were beginning to value forests in different ways, and they began to devise ways to reclaim "extra forest" from villages for the state, in part resulting in the "Three Fixes" (*san ding*) reforms in 1982. This policy significantly expanded the amount of state-owned forests and shrank village-owned lands. With a vast area and few forestry officers, however, these rules could not be rigorously enforced.

12. Ironically this meant that people who kept the forest and wildlife populations in better condition were more likely to be targeted by strict environmental laws than people in places that fully converted forests to fields.

13. The relationship between flooding and upstream human action remains difficult to unravel. Elsewhere in the Himalayas many scholars now challenge what they call the "Himalayan degradation narrative," which blames floods on rural people's livelihood activities such as logging, agriculture, and grazing. These scholars argue that most floods occur because of geological activity and suggest that this narrative is a common instance of states blaming rural people for land degradation (Ives and Messerli 1989; Guthman 1997). Interestingly, in terms of the Yangtze flood, only a few scholars questioned Beijing's explanations or official statistics (see Blaikie and Muldavin 2004; Sauer 1999).

14. By then a number of government departments (including forestry, mining, water management, and agriculture) became increasingly linked to "the environment." Many of these departments' staff were reluctant to follow a new community-based approach for a number of reasons. This included discomfort with undermining their own hard-won social status as urban experts and skepticism that villagers were capable of such decisions, for there was a resurgence of condescending and dismissive urban attitudes toward rural people—i.e., that villagers were "people without brains" (Liu 2000) or could not understand science. Such statements about the incapacity of villages were pervasive and often made by people who had grown up in villages themselves and lived in a city for only a few years. As well, many governmental staff felt a genuine urgency to address environmental problems and dismissed community involvement as requiring too much time and energy.

15. As one of my Kunming colleagues said, this shift to community-based methods reminded him and some of his middle-aged peers of the 1960s and 1970s, when scientists went to villages to "learn from farmers" (*nongmin xuexi*). He conducted workshops on community-based methods for foresters, who were somewhat confused about what to do, so he used Mao-era slogans to explain these methods, seeing this as a return of an old idea, not a completely new one. For accounts of science during the Mao era, see Bray 1986; Han 2008; Schmalzer 2009; Fan 2012.

16. I was one of these critics myself. I collaborated with the ethnobotanist Gary Paul Nabhan and John Tuxill, as well as Elizabeth Drexler, to write *People, Plants and Protected Areas: A Guide to in Situ Management* (1998). We wrote the book for nature reserve staff, providing theories and tools for working with, rather than against, local people. In Kunming ethnobotanists owned this book, in Chinese translation, and my association with Nabhan helped foster my relations with them.

17. It should be noted that coercive conservation manifested quite differently around the world. For example, in Africa guards were given shoot-to-kill orders against poachers of large mammals, whereas, as far as I know, this never happened in Asia or the Americas.

18. See, for example, continuing debates about conservation and social justice between Oates (1999) and Wilshusen et al. (2002).

19. For example, Eric Wolf's *Europe and the People without History* (1982) showed that many societies studied by anthropologists—societies often regarded as traditional, isolated, and "without history"—had been strongly affected by centuries of European colonialism. Since Wolf's publication, others emphasize regions besides Europe, (such as China, Japan, and Africa), which have also been critical in shaping the globe through trade and conquest (Abu-Lughod 1989; Pomeranz 2000).

20. I further explore resistance theories in chapter 3. Many scholars criticize notions of resistance (Abu-Lughod 1990; Ortner 1995; Sparke 2008), but few (e.g. Mahmood 2005) offer much in the way of alternatives.

21. I use the term *accommodation* provisionally, and it should be noted that most resistance scholars assume a subject that is able to resist larger, supposedly hegemonic forces, whereas most Foucauldian scholars see subjectivities as formed in relationship to such forces. Although my own position is closer to Foucault's notions, I am leery of his assumptions of totalizing force, which resemble assumptions of capitalism's totalizing capacities.

22. A number of scholars raise important challenges to theories of governmentality: Gupta and Sharma 2006; Moore 2000; Kipnis 2008; Cepek 2011; McKee 2009; Erazo forthcoming.

23. Some of this research was inspired by cultural studies, which showed that ostensibly passive acts, such as consuming goods (Miller 1995) or media (Spitulnik 1993), were actually creative and active endeavors. Stacy Pigg's (1992, 1996, 2001; Adams and Pigg 2006) insightful work on development and on HIV/AIDS prevention in Nepal explores the difficult work that Nepalis undertake in translating concepts and strategies, both linguistically and socially, across linguistic and cultural borders.

24. Geographer David Harvey (1989) is the most famous source for this description of stages of capitalism, from a Fordist assembly-line mass-production era to that of flexible, just-in-time production of smaller lots.

25. One of the earliest uses of the term *world-making* comes from Lauren Berlant and Michael Warner (1998), who describe the creativity and diversity of queer world-making practices in a deeply heteronormative world.

26. This perspective is related to what is more broadly understood as "emergence," in part inspired by Michael Fischer's work (see special issues in *American Anthropologist*, edited in 2005 by Bill Maurer and *Cultural Anthropology*, edited in 2010 by Kim Fortun, Mike Fortun, and Steven Rubenstein. See also the description of "emerging worlds" by the Anthropology Department at the University of California at Santa Cruz, http://anthro.ucsc.edu/about/emerging-worlds/initiative.html (accessed September 3, 2012). As they argue, anthropologists are moving from a focus on "disappearing cultures" to "emerging worlds," as different groups make their worlds, yet (paraphrasing Marx) not simply as they choose.

27. The term *East Wind* originates from a scene in *The Romance of the Three Kingdoms*, written circa 200 BC, that describes a battle. One army aimed to destroy the other's boats by using burning rafts. After they prepared the rafts, they waited for an east wind to send them toward the boats. The winds blew from the east, the

rafts burned the enemy's boats, and the army emerged victorious. More generally the East Wind refers to preparing oneself as much as one can but knowing that there is always a crucial element of fate (like the East Wind) beyond one's control.

At the time of Mao's coinage, the Soviet Union generously supported China's scientific and industrial development and was often referred to as its "elder brother." By 1958 the USSR had taken the lead against the United States in the Cold War race for military supremacy. After startling the Americans by exploding their own nuclear bomb in 1949, the USSR won the race to launch the world's first satellite, *Sputnik*, in 1957. The satellite was touted as proof of the superiority of "socialist science" over "capitalist science" (Kojevnikov 2008). Whereas China's Great Leap is often analyzed as a domestic event, Mao's invocation reminds us that it was international articulations that helped foster China's initial confidence and subsequent positionings. Although talk of socialist global futures has now largely ceased in China, the term continues to be used, but this time as a metaphor for China itself; the technologies that best represent China's rising global power (satellites and missiles) are still often called "the East Wind."

28. The expression *bamboo curtain* was coined in the West to refer to China's version of the USSR's "iron curtain." In both cases, however, there was actually far more international engagement than the terms imply.

29. In this book I focus particularly on China, but one could also look at the role of places like Cuba, Vietnam, and India, which were also part of revolutionary winds moving through the world.

30. Although we might now think of China as a major superpower, in 1995 it was almost always classified as a "developing country," and my students in Kunming debated whether or not it was a Third World country.

31. As I later mention, China's social influence during the 1960s and 1970s was powerful in many other places, including South America, Africa, and Southeast Asia.

32. There were many connections between African American, Asian American, and Native American activists. Daryl Maeda (2006) shows how American-born Chinese who formed the Red Guards in San Francisco in the 1960s drew heavily on the Black Panthers' theory and style and from Maoist sartorial presentation, wearing armbands and berets. As Daniel Cobb describes in *Native Activism in Cold War America* (2008), Native Americans worked with African Americans in numerous ways, organizing the Poor People's March and sharing strategies such as civil disobedience, sit-ins, and fish-ins. It was only during the 1960s that Native Americans, especially those involved in the Red Power movement, used the concept of "colonialism" to analyze their own histories, as previously it was applied to Africa and Asia, not the United States.

33. Chinese and American study groups were quite different. As Michelle Murphy (2012) reveals, North American groups came up with strategies quite unlike those in China to "take back women's power," such as inventing the vaginal self-exam and framing the right to abortion as an issue for women's freedom and empowerment.

34. The term *Uncle Tom* refers to African American slaves that capitulated to rather than resisted white rule. The term comes from a character in *Uncle Tom's Cabin*, Harriet Beecher Stowe's 1854 book, which was a major force in mobilizing citizens in the United States and England against slavery and precipitating the Civil War. Malcolm X's quote is from a speech in 1963. It is available at http://united blackamerica.com/enemies-foreign-domestic/ (accessed January 17, 2012).

35. While Mao influenced radical black movements, groups affiliated with Martin Luther King Jr. were deeply influenced by another Asian anticolonial with a very different approach: Gandhi (Chabot 2004; Chabot and Duyvendak 2002).

36. This included radical, explicitly anti-imperialist groups like the Weathermen. This group, deeply inspired by China's revolutions, took their name from a Bob Dylan song, "Subterranean Homesick Blues," with its allusion to a shifting wind. Like the Chinese notion of wind, the Weathermen saw social change as a wind that was already blowing around the world and could be enhanced by social action (Rahmani 2006; Rudd 2010). As the term *Third World* implies, these groups were looking not just to China; other major referents included communist Vietnam and Cuba, and some invoked the Soviet Union as an inspiration.

37. During the Mao era China might have seemed isolationist to many in the West, but it was reaching out to groups throughout the world. China devoted tremendous expense and effort to broadcast its messages to the world in film and radio; magazines, books, and newspapers were translated into more than a dozen languages. Overall China was quite successful in creating and cultivating a global audience eager to hear stories of rapid and radical social transformations (Brady 2003; Larkin 1971).

38. China funded, advised, and sent arms to a number of independence movements, including in Algeria, Angola, Congo, Namibia, Mozambique, South Africa, and Rhodesia.

39. China sent twenty-five thousand Chinese laborers to build a railroad, called "the Freedom railway," which stretched almost two thousand kilometers between Tanzania and Zambia (Larkin 1971, 1975; Weinstein 1975; Askew 2002; Monson 2009). In light of China's massive investment in Africa since the early twenty-first century, few commentators recognize this earlier history, and most see China's role in Africa as largely unprecedented. Clapperton Mavhunga (2011), however, warns us against China-centric narratives and argues for attention to how African leaders were not just Cold War pawns but "played" China. For more on African-Chinese relations, see Lee 2010.

40. Richard Wolin's fascinating book *The Wind from the East* (2010) examines the influential legacy of French intellectuals' fascination with Maoism, including such luminaries as Foucault, Sartre, Philippe Sollers, and Julia Kristeva. Wolin shows how landmark texts, such as Foucault's *Discipline and Punish* (1977), were stimulated by interest in Maoist theory and an infatuation with things Chinese.

41. I am referring to a particular formation of environmentalism that arose in the 1960s and 1970s. This is not to say that it was not influenced by antecedents, and I acknowledge the importance of conservation histories, such as Richard

Grove's (1995) argument about the colonial legacies of modern nature and Samuel P. Hays's (1987, 1999) account of the long history of conservation in the United States.

42. In fact the paradigmatic grassroots environmental organization, Greenpeace, was composed mainly of people opposing the Vietnam War who directly and indirectly drew on techniques and philosophies from nonviolent direct action used in the U.S. civil rights movement, inspired by Gandhi's strategies in colonial India.

43. When I later discovered the evaluators' CVs in the archives of WWF-China in Beijing, I could see that they were part of the global development industry; between them they had carried out environmental consultancies in more than thirty countries over nearly twenty years. Their CVs provided a temporal record of the shifting winds of development, when issues like gender and participation, for example, influenced different projects and evaluation criteria.

44. I was in high school, and we raised money by starting the school's first aluminum can recycling program and selling T-shirts that we printed with the quote "Treat the earth well: it was not given to you by your parents, it was loaned to you by your children," incautiously attributing it to Chief Seattle and furthering the association between indigenous peoples and environmentalism.

45. In some locales, farmers and indigenes were seen as overlapping categories, and at other times they were seen as quite distinct and even oppositional.

46. Such encounters were shaped by a particular set of animosities rather than experienced as pan-Asian solidarity. For example, there were particular animosities between Chinese and Indian academics, in part because each believed that their own country was superior: Indians tended to tout their country's tradition of democracy and active public debate against China's "totalitarian state." Chinese were proud of their progress in rural development, which they said was much better than in India in terms of increasing health and living standards and decreasing illiteracy. During the 1990s India had a head start on social forestry programs, and a number of Indian experts were invited to train Chinese foresters and social scientists. At several workshops I attended in Kunming, Indian experts argued that China's population management was "undemocratic," and they cited figures about massively skewed gender ratios in China (many more boys than girls), which they said resulted from a combination of government oppression and Chinese male bias. In response, members of the Chinese audience defensively challenged such figures.

Although Chinese experts often said that "outside knowledge must be adapted to fit Chinese conditions," their way of dealing with Indians was particularly confrontational. Once, a group of Indians explained a participatory technique in which villagers were asked to draw charts in the sand and place pebbles to represent the relative proportion of, say, fuel wood they used to cook food for humans compared to food for pigs. Chinese experts bristled. They said that pebbles might be fine for India, where illiteracy was high, but Chinese villagers would be offended by such methods and would prefer to use paper, pen, and "real numbers." In another gathering, I was placed in the awkward position of "translating" speeches by Indian experts after the Chinese audience said they could not understand their Indian-accented English.

47. Kathy Davis (2007) explores how the world's most influential feminist book on women's bodies (*Our Bodies, Our Selves*) has traveled around the globe. She shows how these international travels not only resulted in differing translations in other languages (as expected) but also affected new editions in the United States itself.

CHAPTER TWO

1. Like Xue Jiru mentioned in the previous chapter, who mainly published as Hseuh Chi-Ju, a romanized name created by the Wade-Giles system (now outdated), Zheng Zuoxin published much work as Cheng Tso-hsin. The Wade-Giles method was nearly hegemonic in Western publications until 1979, when the pinyin system, developed by PRC linguists during the 1950s, started to gradually replace it in much of the world. The conversion from Wade-Giles to pinyin indexes China's growing power and influence, and it was an important event in the year 2000 when the U.S. Library of Congress announced that it would convert millions of Wade-Giles records into pinyin. This project started, coincidentally or not, on October 1, China's National Day. The success of pinyin, created to help people, Chinese and not, learn the language, was itself a form of successful transnational work.

2. The Cultural Revolution took place between 1966 and 1976. A month after Mao's death in 1976, the government launched a campaign to blame the Cultural Revolution on the "Gang of Four," said to be led by Mao's last wife (Jiang Qing) and three others. Over that decade an estimated 750,000 people were persecuted and 35,000 died. The "Destroy the Four Olds" campaign (Old Customs, Old Culture, Old Habits, and Old Ideas) meant that roaming Red Guards, often teenagers, looked for and destroyed evidence of links to China's feudal past (including traditional paintings, ceramics, Confucian objects, and even family genealogies) or imperialism (English-language books were particularly suspect).

3. WWF was started in 1961 in order to raise money for the International Union of Conservation of Nature (IUCN). WWF shared some money with IUCN, but soon became a powerful NGO in its own right. Many of IUCN's and WWF's founders and supporters drew on a particular imperial legacy, wealthy European men who hunted in Africa, later known as the "penitent butchers" after they advocated the animals' protection. They were not interested in "biodiversity" (a term not coined until 1986) but "big game," particularly Africa's "big five:" lion, leopard, elephant, Cape buffalo, and rhinoceros—animals said to present the most challenge and danger to hunters. This history of African wildlife conservation was always gendered and racialized and often as much about "saving" animals for white men to hunt as saving them in and of themselves (Anderson and Grove 1989; Haraway 1989). Yet by the time WWF was created, these dynamics were starting to change, as African decolonization movements gained force. As mentioned in chapter 1, China played an active role in these revolutions. Many Europeans were afraid

that African leaders of newly independent countries would not protect game animals, which were now such a familiar part of European life through zoos, children's books, and stories, that children recognized more African than European animals (Garland 2008). WWF's first major campaign was against rhinoceros poaching, which was remarkably successful in raising money in England, even from the poor. According to one WWF staff member, "We came into being on the backs of rhinos." As I argue in chapter 5, WWF's work in China began on the backs of pandas and elephants.

4. Also in 1979 a small NGO from rural Wisconsin, the International Crane Foundation (ICF), began working in China. They maintained close relationships with WWF. In fact a former ICF representative, James Harkness, later worked for the Ford Foundation and then became the president of WWF-China. I met with Harkness several times and benefited greatly from his knowledge of and insights into environmentalism in China.

5. WWF's mascot was a product of Cold War relations. In 1958 no pandas remained alive in the West, but an American zoo had somehow arranged to purchase one, which arrived at a U.S. customs office. Because of the American trade embargo against China, the panda, Chi Chi, was refused entry and stuck in international limbo. In a hastily arranged deal, the London Zoo purchased Chi Chi, and she immediately become the zoo's star attraction and one of England's most well-loved animals. Chi Chi was the basis of Peter Scott's design for the World Wildlife Fund logo. In July 1972 Chi Chi died and was publicly mourned. Her funeral was one of England's most well-attended displays of national grief for nonhuman animals (Morris 1966).

6. Schaller was eager to work with pandas, especially as his field studies on the tiger, a major project between the Indian government and WWF, were recently derailed. As part of India's growing nationalism, officials in New Delhi refused to issue visas to Schaller, hoping to replace him with Indian biologists (Lewis 2004). At this time, India was quite leery about being used as a pawn in America's Cold War struggles with the Soviet Union. Indian officials heard rumors that American conservationists affiliated with the U.S. military were studying birds in India as potential agents of biological warfare, especially those that migrated to the Soviet Union (Lewis 2004). Although Schaller found research in China challenging and quite bureaucratic, he was grateful that the Chinese were opening doors to foreigners just as the Indians were closing them.

7. It is important to remember that in the 1980s relatively few Western environmentalists were interested in China. In their eyes, China was mainly an overpopulated nation with little, other than pandas, worthy of ecological protection (Smil 1984).

8. A strong pride in disproving foreigners and possessing something rare and valuable was evident in reactions to the prince's trip. For example, one writer stated, "In the past, some foreign scholars asserted that 'China has no plants of the camphol plant family' or 'China has no tropical rain forests.' However, the discovery of the wang hsie [*Parashorea chinensis*, a Diptocarp often seen as an exemplary tropical

rain forest tree] not only overthrows such assertions, it also proves that China has tropical rain forests in [the] real sense," http://www.kepu.net.cn/english/banna/tropic/tro414.html (accessed February 3, 2012).

9. As mentioned in chapter 1, the term *slash and burn* has a negative connotation. In this book, I use that term or *swidden cultivation* or *shifting cultivation*, depending on the context. The latter terms can be neutral or even positive. Therefore, when talking from the perspective of those trying to stop it as a negative practice, I use *slash and burn*. When talking from a neutral or positive position, I use *swidden cultivation* or *shifting cultivation*. I describe the history and politics of English and Chinese terminology for this practice in chapter 4.

10. I draw on Stuart Hall's concept of articulation (inspired by Antonio Gramsci), which is used by a number of scholars, especially those who explore how social conjunctures, such as racial and ethnic formations, come into being in particular times and places (Clifford 2001; Li 2000; García 2005). It is an anti-essentialist position, for it stresses contingency and social practice. Like Tania Li (2000), for example, I find the concept helpful in understanding why and how some "tribal" groups link with global indigeneity while others do not. One potential limitation of the notion of articulation, however, is that it may foreground initial connections at the expense of understanding the ongoing efforts needed for its continuation, as well as its dynamic transformation.

11. One of the key elements of the PRC's post-1989 foreign policy was a mandate for Chinese officials to *qiu tong cun yi* with their foreign counterparts, that is, "looking for things in common and letting disputed points lie" (Brady 2003: 218). This approach may have influenced WWF's negotiations over nature conservation.

12. Chongqing, previously known in the West as Chungking, is now officially a municipality, a major city almost equal in status to that of a province.

13. These were Beijing University, Tsinghua University, and Nankai University, which were merged in Kunming for eight years and were known as National Southwestern Associated University.

14. Some scholars say that for Chinese scientists, 1949 to 1957 was the PRC's "honeymoon period." We will learn much about debates over conservation with future work by historian Elena Songster. Also see Yao 1989; Schmalzer 2008; Neushul and Wang 2000.

15. The Anti-Rightist campaign came after the "One Hundred Flowers" campaign that promised to let "one hundred schools of thought contend, and one hundred flowers bloom," supposedly indicating the government's openness toward critique and free expression.

16. The term *old stinking nines* was not a new one but an epithet dating back to the Yuan Dynasty to designate intellectuals. The persecution of intellectuals during the Cultural Revolution included people such as botanist Cai Xitao, who helped create the Xishuangbanna Tropical Botanical Garden and was later declared one of Yunnan's most influential scientists (Li 2010). It is a remarkable that despite years of persecution, Cai and many other surviving scientists returned to carry out years of cutting-edge work.

17. Through great efforts, China circumvented the embargo with third parties and barter. Starting in 1954 Chinese officials obtained rubber by trading Chinese-made industrial equipment for 200,000 tons a year of rice from Burma (present-day Myanmar), which was then exchanged for rubber in Ceylon (present-day Sri Lanka; Fung 1972). As well, Indonesia and other sympathetic countries covertly sold rubber to China (Lu 2008).

18. In 1962, to their chagrin, Chinese officials had to import grain for the first time since 1949, buying wheat from Canada and Australia (Fung 1972). Under Mao, China quickly became a net grain exporter, sending mass quantities to the Soviet Union to pay for industrial equipment. Although a source of pride, such large-scale exports were later thought to have added to China's famines.

19. During my first trip to China, in 1995, many people, including bicycle repairmen and clothing vendors, told me that science was the main driver behind national standing in the global ecumene. I also noted that officials and academics frequently talked disparagingly about farmers' ability to understand science.

20. Although most publications in China refer to this plant as *Eupatorium*, it is likely *Chromolaena odorata*. In Southeast Asia this plant is called "airplane weed," said to have been dropped by Americans or Japanese as biological warfare to reduce agricultural productivity. It spread quickly, intentionally or not, during World War II (McFadyen 2003).

21. As I describe in detail in chapter 4, these views of slash and burn were not hegemonic. During the 1980s and 1990s a handful of ethnobotanists and anthropologists helped to position slash and burn as "shifting cultivation" and show it in a positive light. Their efforts, in part, led to WWF's changing interpretations of its project and the rise of an "indigenous space."

22. This sentiment and expression was still strong into the early twenty-first century, when I heard similar statements from several local officials. Clem Tisdell reports that a leader told him that, faced with the choice of conservation or feeding the people, they would always choose feeding the people (in Lehane 1993: 2).

23. In part this shift occurred because vast areas of forests were redesignated in 1982, from land earmarked for supplying farmers' subsistence needs (i.e., for shifting agriculture and firewood collection) to the property of the State Forestry Department. Had WWF come to China in the early 1970s, there would have been little support for nature conservation projects other than those to support panda bears.

24. Some Chinese officials knew that WWF was based in Gland, Switzerland. A number of WWF-China's representatives, however, and the main Yunnan project workers were American, so this may have led to such a misunderstanding.

25. Most universities reopened only in 1979, after a thirteen-year hiatus.

26. Wang and Xue, in turn, relied on work by Chinese and Western botanists, who began in earnest only in the early twentieth century to document Yunnan's animals and plants in relationship to Linnaean taxonomy with Latin names. See Mueggler (2011) for fascinating accounts of this history.

27. Scientists continue to value these old books and articles as critical references, especially for scientific papers trying to declare a new species. These books

are often quite difficult to obtain but necessary for scientific validation. Most of these old texts are in English, but some are in French or German. For example, one paper from 2007 by Ye and others cites the following: Anderson's 1878 *Reptiles and Amphibians*, Boettger's 1885 *Materialien zur Herpetologischen Fauna von China*, and Boulenger's 1920 article "A Monograph of the South Asian, Papuan, Melanesian and Australian Frogs of the Genus *Rana*" in *Records of the Indian Museum (Calcutta)*. I met some scientists in Kunming who spent years trying to obtain copies of such texts for their research.

28. Such fears remained deep; for example, a 1998 poll in Germany called the "Worry Index" indicated that people's greatest anxiety was the destruction of the rain forest (Stott 1999).

29. Although *The Population Bomb* was attributed only to Paul Ehrlich, an article (published more than twenty years later) describes Anne's extensive contributions to the book (Ehrlich and Ehrlich 2009).

30. China's official position on the role of population has changed. Beginning in the 1930s a number of prominent Chinese officials suggested that socialist reconstruction would be assisted by having more people (White 1994). Susan Greenhalgh's work shows the complexity of the debate (2003, 2008, 2010), but, put simply, such officials dismissed the idea that population inevitably strained resources as capitalist rhetoric and as Malthusian and bourgeois concepts that did not apply in socialist contexts. Their use of the term *Malthusian* refers to theories developed by Reverend Robert Malthus in the nineteenth century that human populations will increase more rapidly than food supplies, an idea that quickly gained traction and led some (who often held a social Darwinist or "survival of the fittest" position) to reduce support for social welfare. Since the end of the Mao era, however, leaders have reversed their position and describe overpopulation (*renkou tai duo*) as one of China's main challenges (Boland 2000).

31. MacKinnon articulated a widespread sentiment when creating a poster for a display at the Xishuangbanna Nature Reserve Bureau. He wrote, "In China, overpopulation is most serious in areas of minority nationalities where birth control programmes do not apply. As the land degrades through overexploitation by the ever-increasing population, bigger fields are needed to grow food and more children borned *[sic]* to provide labour; a vicious cycle goes on" (Cheung and MacKinnon 1991: 67).

32. Tang Xiyang became an important figure in fostering environmentalism in China, starting a journal called *Great Nature* in 1980 and writing a best-selling book, *Green World Tour*, based on eight months of travels around the world to nature reserves with his American wife, Marcia Bliss Marks. His book stimulated much public interest in nature, inspiring many Chinese citizens to send money to create nature reserves. Tang remains an active speaker, writer, and advocate well into his seventies. His position that rural villagers were a major cause of ecological destruction paralleled another prominent environmentalist of his generation, Liang Congjie, who started China's first environmental NGO in 1994. When anthropologist Robert Weller interviewed Liang, searching for "indigenous views of

Nature" (here *indigenous* means Chinese, or non-Western), Liang answered that not only did villagers not have valuable environmental knowledge, but that "China's biggest problem ... was the environmental ignorance of the peasantry" (paraphrased by Weller 2006: 129).

33. It should be noted that in China from the 1950s to the 1970s, descriptions such as those used by Tang Xiyang and John MacKinnon would be considered elitist and condescending and possibly labeled as "Han chauvinism" (*da hanzu zhuyi*) or "splittism" (*fen lei*, driving a wedge between the hoped-for unity between all ethnic groups). The government continues to put much effort and funding into "minority work" (*minzu gongzuo*) to reduce ethnic tensions and foster unity.

34. Political ecologists have long pointed out problems with this perspective, which ignores how rural lives are constrained by the political economy (Blaikie and Brookfield 1987). To be fair, a number of conservationists, such as Norman Myers (1984), recognized the importance of larger forces and saw rural peoples as pawns of wealthy landowners, corporate resource extraction, and so forth.

35. Ironically, some of these same scientists training farmers in the 1980s had earlier been sent "down to the countryside" to learn from farmers.

36. In the year 2000, I discovered Yang Bilun's heterodoxy in a conversation about the Chinese term for "culture" (*wenhua*). In China wenhua can be described as high or low, present or absent. Wenhua specifically refers to the degree of formal education, so that someone without any schooling might say, "I don't have any culture" (*Wo meiyou wenhua*). Yang described the term *wenhua* as Han-biased. Whenever his Kunming-based scientist peers traveled (usually reluctantly) to rural villages, they would always ask villagers about their level of wenhua. This offended Yang, who instead asked residents specifically about their level of Han-style formal education (*Hanzu jiaoyu shuiping*) to avoid conflating culture with academic opportunities.

37. As an EU report stated, "The project proposal was written without any form of rigorous identification mission and without the participation of villagers or key officials in China" (Newman and Seibert 1995: 36).

38. Bentley did not explain why lowering crop yields might lead to a larger family size, and many of his colleagues elsewhere in the world made similar unsubstantiated claims (O'Brien 2002). This was true not only for slash and burn but also for overgrazing and desertification. Only in the mid-1990s were these assertions challenged, such as in Melissa Leach and Robin Mearns's landmark anthology, *The Lie of the Land: Challenging Received Wisdom on the African Environment* (1996).

39. As Kathy White reminded me, this situation occurs in other places, such as Nepal, where foreigners can't read the work of local people and thus tend to create self-referential Anglophone realms of knowledge. This problem is less common in Spanish- or French-speaking countries, where Anglophone expatriates are more likely to know the language. Language politics were quite important in Yunnan's transnational work. I do not suggest that Bentley or MacKinnon should have read or spoken Chinese, and few short-term foreign workers could be expected to read Chinese, which requires years of dedicated effort. Foreign NGOs therefore relied

heavily on bilingual Chinese experts. English was by far the most common intermediate language at international conservation organizations in China—the lingua franca, as it were, in China. Even organizations staffed primarily by German, Dutch, or Japanese nationals wanted English speakers, and they rarely hired domestic staff who knew only Chinese and their own non-English language. To be fair, the expatriate staff were always at least bilingual, English being one of their languages. In fact Dutch and German staff confided to me that they liked having their own private language, discussing sensitive issues without involving their Chinese colleagues, or handwriting messages in Dutch or German, worried that email might be monitored. Overall monolingual Chinese scholars found it difficult to be hired by foreign NGOs or establish themselves internationally, as Chinese texts were rarely read beyond China. A very small percentage of Chinese studies were translated into English, although a number of Chinese scientific journals now mandate that the articles provide an English-language abstract, and some even require that the entire article be written in English. Some experts paid substantial fees to translate their writings into English or got their institution to pay, and a few groups like the Ford Foundation paid for a handful of translations of books that squared with their perspective, such as those endorsing rural social justice (e.g., Yin 2001; Gao 1999; Guo 1998).

40. At this time, experts could not simply send in an application for a passport to Beijing but first needed to get permission from their work unit (*dan wei*). I heard of several cases in which an individual did not have good relations with the relevant authorities and found it difficult or impossible to obtain a passport. For example, one scientist I knew waited several years for a passport, until one particularly obstructive individual retired from the post. In several cases, opportunities arose just before applications were due, so that an invitation to attend a conference in Thailand, say, might appear only several weeks before the event itself, leaving insufficient time for those without a passport to apply for one.

41. It was not only WWF. Kunming became the host to a number of international organizations and conferences, such as Alternatives to Slash and Burn in 1995, which offered opportunities for Chinese scholars who attended, presented papers, and socialized with their peers from around the world.

42. Yet, as we will see in chapter 4, several Chinese experts in Kunming challenged these narratives and built networks of allies (often based outside of China, but they also needed domestic supporters) to foster their work. These few experts were the notable exceptions to the rule.

43. Just to reiterate a few of the differences: MacKinnon wished for stronger birth control and antihunting campaigns against ethnic minority groups (but Chinese officials were leery of creating social conflict) and reducing rubber plantations (but WWF staff were worried about offending Chinese officials). Chinese staff from the Ministry of Agriculture were often skeptical of agroforestry and wanted to promote what they called modern cash cropping instead, and a number of local-level officials were opposed to strict conservation laws.

CHAPTER THREE

1. Interestingly, very few people spoke Jinuo. Some said they were members of another group, which they called Benzu, but the government did not recognize them as distinct, and they were given Jinuo status. From the project's beginning, staff from WWF and the nature reserve were not sure what to think of Xiao Long residents, wondering if they were really Han majority or Benzu or Jinuo. After a reserve staff member attended a Xiao Long wedding, he told me that their singing and dancing "proved they really were an ethnic minority group, because the Han [majority] don't sing and dance together."

2. Although Scott more frequently points to these less often noticed, subtle, and individual actions, most influential studies of indigenous resistance in places like Latin America point to direct confrontations, such as mass demonstrations with placards, marches, and road blocks (Jackson and Warren 2005). More recently anthropologist Maria Elena García (2005) has shown that in countries like Peru, long seen as lacking an indigenous movement, indigenous engagements are still vibrant and important, and Juliet Erazo (forthcoming) shows that in Ecuador indigenous groups are not just "resisting" governmental plans, but their leaders reach out and enact everyday forms of governmentalization among their own populations.

3. Unlike most scholars of *guanxi xue*, I stress its continuing significance for rural people in establishing connections with others, including urban residents who often possess greater social status, power, and wealth. I argue that rural citizens often try to increase connections with powerful outsiders, who may or may not be state officials, on behalf of themselves and sometimes their village. This differs from the argument of one influential scholar of China, Mayfair Yang (1994: 172), who interprets individuals' use of guanxi as a kind of "social resistance against total state saturation," thus mirroring some assumptions about state-society relations found in James Scott's work. I agree that the use of guanxi creates extra-official pathways in lieu of official protocol, but I do not see this as a matter of resistance; I see it more as engaging than resisting.

4. WWF left Xishuangbanna and moved its projects to Northwest Yunnan. These two regions began to compete for domestic and international attention and funds. WWF's presence helped Northwest Yunnan displace Banna as the province's new center for global ecological attention, and the Northwest's influence was magnified by a series of major projects by The Nature Conservancy and then by Conservation International, starting in the late 1990s. Some colleagues suggested that by the time I arrived for fieldwork in 2000, Banna was like a Chinese fried dough stick (*youtiao*) with all the oil squeezed out, which is to say, used up.

5. GTZ is a German bilateral organization, now known as GIZ (Deutsche Gesellschaft für Internationale Zusammenarbeit), substituting "international" for "technical."

6. In dramatic contrast to these studies by politically progressive scholars, the progressive public in the Global North largely understood environmentalism as a

positive force. They tended to understand it as a grassroots movement that pressured governments and companies to preserve wildlands from development, clean up pollution in places like Love Canal, stop large-scale forest clear-cutting, and save whales from industrial harvest. North American media almost always celebrated international conservation efforts and rarely mentioned any negative social consequences. The British media, in contrast, has carried a number of critical stories, for example, exposing hardships that Western conservation activity imposes on rural Africans. These stories are detailed in Raymond Bonner's *At the Hand of Man* (1993), a biting exposé of how powerful Westerners and their conservation efforts cause difficulties for many Africans.

7. This was especially true for Africa and Asia (Neumann 1998; Guha 1990), but it also applies to Latin America and North America (Klooster 2000; Spence 2000; Jacoby 2003).

8. For my first several weeks living in Xiao Long, some wondered if I might be the son of Jack Bentley, WWF's agroforester. They thought I might have returned to evaluate my father's efforts and consider bringing another project. Although people eventually learned that this was not true, I was seen as a potentially powerful outsider, depending on my connections with Chinese cadres, businessmen, and foreign organizations. They found it puzzling that an anthropologist would want to visit them. A few people had heard of anthropologists, as they had relatives in nearby Jinuo Mountain, said to be the cultural center of the Jinuo ethnic minority group, the last group given official recognition by the government in 1979, where some older Japanese anthropologists studied, according to villagers, "old superstitious rituals." However, they didn't think of themselves as possessing much of anthropological interest, as they didn't speak non-Chinese languages or know much about distinctive customs (*fengsu xiguan*). Instead, while most used my Chinese name (He Wei, for Hathaway), others jokingly called me "the reporter" (*jizhe*), for I frequently conducted interviews and wrote down observations in a small notebook.

9. There are different lineages to resistance. Subaltern studies drew less on Scott and more on the work of the Italian intellectual Antonio Gramsci, famous for his notions of hegemony as an always partial and incomplete project (e.g., Guha and Spivak 1988). They were particularly interested in recovering a subaltern "voice" and offering a historical account of Indian anti-imperial efforts that highlighted the role of subalterns, as opposed to dominant accounts written by colonial and nationalist historians that often privileged the role of elites. Scott argues against a strong version of Gramscian hegemony that doesn't seem to allow room for dissent and differing perspectives.

10. Bernstein (1990), a long-time editor of the *Journal of Peasant Studies*, remarked that such tropes of white-hat peasants and black-hat officials were common in manuscripts submitted to his journal.

11. Especially in Latin America scholarly discussions of indigenous people's agency are quite often described as resistance to neoliberalism and globalization (e.g., Renique 2009; Fenelon and Hall 2008; Bargh 2007). As Peter Wilshusen (2010: 771) argues, however, "an exclusive focus on resistance removes from view

those settings where neoliberal policies and programs are partially or wholly assimilated and the everyday processes by which such accommodations are constructed." Instead Wilshusen's notion of "creative accommodation" argues that, "unlike resistance movements such as the Zapatista uprising in the Mexican state of Chiapas, practices of accommodation emerge slowly and incrementally over time in the course of everyday interactions" (770). My interest in the art of engagement includes an exploration of resistance and accommodation, as well as the ways these encounters transform each, so that policies are not only accommodated but transformed through cumulative practices.

12. There are a similar set of assumptions around Western notions of activism, which are linked to a certain notion of progressive politics.

13. It was not always like this. Older men and women in Banna talked about the 1940s and 1950s, when the People's Liberation Army reduced banditry but also "made them poor" by destroying their opium-based economy. This did not happen overnight, and opium continued to be grown into the 1960s in some places.

14. Erik Mueggler's eloquent book *The Age of Wild Ghosts* (2001) reveals the long history of how one upland Yunnan ethnic minority group, the Yi, created complex social institutions to host visiting officials.

15. We know little, for example, about uplanders' religious and economic worlds, in part because scholars have often simply described uplanders' beliefs as a generic "animism," which eliminates further inquiries. Scholars often assumed the uplanders maintained a primitive self-sufficiency rather than investigate their history of trade and interrelationships. It was also difficult to carry out this research because of the legacy of officially recognized terms for the various groups. One of the upland groups (the Jinuo) was given official status only in 1979, so there are few studies of this group before this time, as almost all studies are generated using state-sanctioned ethnic categories. We also know little about connections between groups, as Chinese studies are almost always done of single groups and accounts of relationships between groups tend to be quite coarse. Most accounts written during the Mao era described how ethnic minority groups were exploited; most accounts written during the post-Mao era tend to focus on cooperation. Some of the richest historical and ethnographic accounts are found in work by Giersch (2006), Wiens (1954), Borchert (2008), Sturgeon (2005), and Hansen (1999, 2005). Borchert (2010) argues that Western scholarship on ethnic minorities in China often presents the relationship as having a Manichaean dynamic, with Han oppressors and ethnic minority resisters, following the "David and Goliath" pattern described by Harold Bernstein (1990).

16. This was true until the 1960s, when China began mosquito-eradication campaigns and made medicines, including effective antimalaria drugs. Such medicines were an important part of China's Cold War efforts in Africa, and one, *qinghaosu*, from the *Artemesia* (mugwort) plant, still remains more effective than quinine (Li 2011).

17. "Sent down youth" or "intellectual youth" (*zhiqing*) refers to the young adults (many in their teens) who participated in the Down to the Countryside Movement

in the 1960s and 1970s. They were sent from cities "down" to rural areas to work alongside and learn from peasants. Many zhiqing were sent to other parts of Banna, often working on rubber and tea plantations. In 1979, after years of agitation, many received permission to return to Beijing, Shanghai, or Kunming (Yang 2009). In the 1990s some former zhiqing visited the villages where they had been based, and some set up business connections or helped fund better schools.

18. It seemed that other animals, like the small forest deer (*jizi*, a muntjac; *Muntiacus muntjac*), were still fair game. They were plentiful and often came into the rice paddy to eat the growing plant stalks and were seen as a pest, as they were more threatening than those animals that ate only the mature rice seeds. The seed-eating animals would typically come only during a limited time frame (normally during the last weeks of ripening, when families tried to post guards to protect the crop), but the stalk-eating animals could come at any time, and rice and corn could not always create seeds after being grazed.

19. Despite having its headquarters in Geneva, WWF was seen as an American organization by many lower-ranking officials and villagers because its head and Bentley were both American.

20. Many studies of resistance present two sides, each relatively fixed. Ethnographic studies of international development projects often underemphasize the role of the state and instead stress the global and local. When studies present three sides (often global, local, and state), global groups (such as capitalists or environmentalists) are often regarded as working in tandem with states to impose regimes on local peoples. Yet if we look closely at WWF staff, we can see that not all agents of globalized social formations, such as Bentley, worked in collaboration with the state. More importantly, I show how such divisions themselves might be questioned.

21. The Philippines, a country with an Anglophone elite and relatively easy accessibility to groups from the Global North, has been a key site for new experiments in development. Places like China, in comparison, have far greater restrictions on efforts by international NGOs and missionary groups.

22. Later, in the mid-1990s, WWF (1996: 9) opposed relocation, mainly because of changing international understandings of local rights and also due to efforts by Chinese experts like Yu Xiaogang and Xu Jianchu, who criticized these policies and advocated for greater recognition of indigenous knowledge and rural rights. I discuss this in chapter 4.

23. Bentley was passionate about raising fruit trees and horses on his Montana ranch. According to him, his role as a village compatriot was strengthened when his wife, Marsha, accompanied him on a visit. Bentley said that she, raised on a farm as he was, helped village women with tasks such as carrying water and wood (Bentley 1993: 32).

24. As an American researcher, my own status was always in question. During a trip in the spring of 1999, I was asked why the United States intentionally bombed the Chinese embassy in Yugoslavia. In 2000 a U.S. Navy surveillance plane landed on Chinese territory after it collided with and destroyed a Chinese jet. Although I

was not treated badly after these incidents, they were continual reminders of historic antagonisms. At the beginning of fieldwork, one old woman took Yang Bilun aside and whispered to him, "Before I ask that American something, I want to know if we [China] are against the United States right now." He laughed out loud, thinking that it had been almost forty years since Nixon's visit meant the United States was classified as a "friend," not an "enemy," but we both recognized that for much of her life, she was told that the United States was China's enemy. When I got to know her better and we sat down with some of her sons to share a bowl of bamboo rat soup, she said she had participated in numerous anti-America campaigns and that her friend, Old He, had lost his eye to an American bullet in Korea.

25. I heard of numerous examples in which suspicions went both ways. Like Bentley, other NGO staff in Kunming (at groups like Oxfam, Doctors without Borders, The Nature Conservancy, and Save the Children Foundation) said that the Chinese state interfered with their projects; a few believed that "government spies" followed them. In 1995, after returning from an awards ceremony and feast in Beijing, Bentley said that the government was "up to its old tricks." Bentley lay in bed violently ill. "They [Chinese government officials] were showing me that they are still in control. They gave me this award, and then they tried to poison me. Maybe not to kill me, but just to teach me a lesson." Whether or not he was the victim of an intentional or accidental food poisoning (and I tend to believe the latter), his interpretation resonated with stories I heard from other expatriates, including anthropologists.

26. There were parallels between Xiao Long and the college. Village residents, relatively poor and increasingly isolated, were worried about becoming irrelevant. The college had always had a relatively low status and was now threatened by shrinking state funding and greater competition from newly emerging private colleges. Previously, high school students were assigned to a college by the government based on their test scores, but now colleges could attract students directly to their school. In each case, their lives were shaped by environmentalism. In Xiao Long the winds meant more restrictions on their everyday livelihoods. The college worried after logging was banned, and established a new program requiring that top students be trained in tourism. It then created a major in conservation biology, the first in China, as it became a key training ground for environmental protection. Xiao Long had fewer resources and less room for maneuver. Both college and village knew that reaching out to powerful outsiders (foreign and domestic) could bring an increase in status, wealth, and opportunity, but it also created more risk.

27. Bentley was not able to produce any statistics from this site, as WWF did not provide funding or the labor needed for creating comparative sites to compare yields. Like many organizations, they trusted the glowing reports of agroforestry.

28. This plot did not represent all of the landholdings for these ten families, as they all had other land elsewhere.

29. The leader at the time, who was quite popular, first purchased a tractor, dubbed a "Donald Duck tractor" (*tang laoya tuolaji*), for its awkward waddling motion.

30. WWF staff purchased some existing fishponds, removed the carp, and experimented with fish more common in African development circles, like tilapia.

31. They would serve him noodles or potatoes, special foods in comparison to a rice-based dish for every meal.

32. This was part of a generalized and long-standing Chinese practice, by which immigrants create "hometown associations" when they move to other regions or countries (Liu 1998).

33. On the other hand, some groups (such as Uyghurs and Tibetans) may desire a degree of political autonomy. For example, some Uyghurs connect with a larger diaspora, mainly based in Germany and the United States, who have worked since the 1940s to create their own nation, East Turkmenistan, partially in Chinese territory.

34. Ann Anagnost (1997) and Dee Mack Williams (2002: 88) provide interesting accounts, during and after Mao-era China, of the benefits gained by being seen by state officials as an exemplary individual or village. Anagnost, for example, describes the "civilized village campaign" and shows the many efforts that some places did put forth to contend for this title, knowing that it could provide positive attention from state officials.

CHAPTER FOUR

1. Such a question has obvious parallels with other transformative movements, such as those that repositioned epithets such as *queer*, *black*, and *Quaker* into self-described social categories that foster political alliance. *Tuzhu ren* will not necessarily gain traction in China. In nearby Taiwan, the well-organized Alliance of Taiwan Aborigines, begun in 1984, urged the United Nations to drop tuzhu ren because it was widely regarded as connoting "primitive" people of "low cultural level," and instead use *yuanzu minzu* (indigenous peoples), which connotes temporal precedence (Kingsbury 1995: 357).

2. My use of the term *indigenous space* is related to, but different from, Tania Li's (2000) notion of the "tribal slot." Li's term refers to the externally imposed expectations that a tribal or indigenous group is self-sufficient, not market-oriented, isolated from dominant society, "traditional," and environmentally inclined. Li builds on Michel-Rolph Trouillot's (1991) "savage slot," the West's long-existing fascination with "the savage," which, like "the Oriental" (Said 1978), is surprisingly enduring. Li points out that there are potentially many groups that could hope to articulate with indigenous rights NGOs, but these expectations powerfully limit which groups may and may not be deemed indigenous. Other scholars have revealed similar difficulties, in which a narrow range of expectations about how indigenous people talk, dress, and act presents increased challenges in winning court battles in the United States, Canada, Australia, and Brazil (Clifford 1988; Hamilton 2008; Povinelli 2002; Ramos 1998) and dealing with environmentalists (Gra-

ham 2002). My perspective on globalized indigeneity builds on insights by Ronald Niezen (2000, 2003) and Anna Tsing (1999, 2007).

3. Although a number of scholars debate how well concepts like "race" apply in China (Dikötter 1992), my conceptual understanding of changing ethnic politics there is indebted to Michael Omi and Howard Winant's groundbreaking book, *Racial Formation in the United States* (1986). Omi and Winant challenge the idea of a singular and stable racial configuration (such as "the" American racial structure) and provide a rich and fascinating account of how contending "racial projects" are at work, shaping new kinds of racial formations. This sensibility inspires my efforts to think about how winds are historically grounded movements of social change and how "work" is carried out by many actors with diverging aims. One small criticism of their book is that it lacks inquiry into how dynamics beyond America's borders played a role in these changing racial projects; however, it should be admitted that most studies focus solely within the borders of one nation.

4. Of course, the term *savage* also carries many pejorative connotations, and the status of nobility is always precarious and easily lost if any action contradicts the high expectations of righteous behavior. As Emily McKee reminded me, it is this double meaning that makes working with notions of the noble savage both useful and dangerous for environmentalists and indigenous rights advocates.

5. In some cases, missionaries created written languages for such groups, but such systems were often not recognized by the Chinese state, which stopped publications in those language.

6. We shouldn't assume a North American or European model, which often views "ethnic minorities" as immigrants who have only recently gained a modicum of political power. It is worth noting that two groups that are now considered ethnic minorities, the Mongols and Manchu, controlled China for centuries. The Mongols (under Kublai and Genghis Khan) ruled one of the world's largest empires during the Yuan Dynasty (1271–1368). The Manchu controlled the Chinese Empire for nearly three centuries during the Qing Dynasty (1644–1911).

7. The one example of a group that did try to connect with indigenous networks but has largely abandoned such efforts is an Uyghur support group (Uyghur Human Rights Project 2009) in the United States.

8. At the same time, elsewhere in the world, such as North America, there were many groups involved in legal battles to gain government recognition as First Nations or Native Americans and to expand land rights.

9. This was not the case only in China, and groups around the world, such as the Maasai and San in Africa, were unaware of this transnational category and engaged with it only after persistent efforts by indigenous rights advocates. In part, they were not aware of it because the governments of Kenya and Botswana refused indigeneity (Hodgson 2002; Sylvain 2002).

10. In 2000 the World Bank withdrew support from the massive Qinghai Anti-Poverty Project in western China, in part on the grounds that some of the affected peoples were indigenous (Bottelier 2001). This withdrawal was not initiated by the World Bank itself, nor by grassroots protest in China, which would have been

quickly stopped by police. Rather the protest was mounted by Tibetan rights advocates who argued that the project contradicted Bank policy on indigenous peoples. We should not imagine the Bank as a major proponent of indigenous rights, given that it initially supported the project. What it does mean is that the Bank is now somewhat accountable to publicly made accusations that it does not follow its own policies on indigenous people. In a number of other cases in China, Bank officials have argued that they considered special plans for indigenous people but declared that the "the majority of beneficiaries are minorities," implying that as "minorities" they are not "indigenous" (World Bank 2008). Thus in China projects by the World Bank or Asian Development Bank hinge on whether particular groups are classified as "indigenous" or "minorities," with stronger rights allocated only to the former. Likewise these cases show that the World Bank and other international organizations do not have the ability or the desire to impose the category of indigenous people on China or any other country, but they can be vulnerable to lobbying. As China becomes increasingly wealthy, it is less dependent on World Bank funding. In this instance, after the Bank withdrew from the Qinghai Anti-Poverty Project, China funded the project itself.

11. Throughout the book I use pseudonyms. In this chapter, however, I use the real names of Pei, Xu, and Yu; they are public figures and their work is so distinctive that attempts to disguise their identity would be fruitless. Although I draw on conversations with them over many years, I mainly refer to publically available materials.

12. As detailed elsewhere, the Dai kingdom's connection to imperial China was complex, and they simultaneously gave tribute to imperial China, Burma, and Siam (Giersch 2006).

13. As I describe later, the term *shensheng* (often translated as "sacred") was used to designate subjects such as ancient Chinese beliefs in the Five Sacred Mountains, but it was not a well-known term. *Mixin,* on the other hand, is an everyday term, especially used by rural minorities who are often self-conscious about beliefs and practices that for decades were subjected to castigation or punished (also see Chao 1996).

14. Whereas *fengjian* refers to a Chinese Marxist notion of feudalism as a problematic ideology that needs to be routed out, *mixin* has an older legacy. As Mayfair Yang (2011: 6) argues, *mixin* was already a pejorative term used by nineteenth-century missionaries, "a catch-all category that included practices that Protestantism had rejected, such as magic, shamanism, deity and ancestor worship, exorcism, demon and ghost worship, polytheism, ritual healing, divination, communicating with the dead, etc. All of these practices were important aspects of traditional Chinese religiosity, but, in the move towards Westernization, they came to be regarded by Chinese elites and officials as 'feudal superstitions' that had to be swept away." I would agree with all of this except that such elites and officials did not necessarily promote a "move towards Westernization" per se but did promote a move toward modernizing China, which was not necessarily the same thing.

15. Campaigns against superstition were also carried out before Mao's time but within a very different and limited social context (Nedostup 2009).

16. Although I wanted to uncover the history of this forest patch, such as who protected it, when and why, I was unable to find people who knew precise details. Erik Mueggler (2001) describes some of these assaults on trees that were regarded as emblematic of feudal superstition in Yunnan.

17. In contrast, a decade earlier there was little mention in China about environmental concerns, especially in terms of the value of natural forests in and of themselves.

18. Yin Shaoting's book *A Highly Controversial Cultural-Ecological System: Studies on Swidden Cultivation in Yunnan* (1991) shows that he was one of the earliest in Yunnan to challenge derogatory descriptions of slash and burn. It was difficult to get much domestic support for his perspective, but he got help from abroad. While many tend to emphasize China's links with the West, people like Yin, Pei, Yu, and Xu were active participants in Asian regional networks. Yin was in Japan as a visiting fellow from 1990 to 1996, a guest researcher and professor respectively at the National Museum of Ethnology, Kyoto University and the Tokyo University of Foreign Language in Japan, There, some of his articles and books (e.g., Research on Yunnan Swidden Agriculture in 1996 and "The Swidden Agriculture of the Jinuo Nationality" in 1992) were translated into Japanese and English. In Japan there is strong public and scholarly interest in the lifeways of Yunnan's ethnic minorities, some of whom are regarded as Japanese ancestors (Oakes 1999). As Yin's heterodox views became less controversial in China, he returned to Yunnan, where he enjoyed prestige and influence, garnering support from groups like the Ford Foundation for creating local museums and translating one of his books into English. He later became the chair of the anthropology department at Yunnan's preeminent university.

19. In China the official term for rural citizen is *nongmin*, often translated into English as "peasant" or "farmer." In general, authors who use the English term *peasant* often mean to imply less favorable connotations, describing such people as lacking in initiative or knowledge, whereas farmers are more often regarded as hardworking and independent.

20. See also Dee Mack Williams's ethnography, *Beyond Great Walls* (2002), for examples of how dominant Chinese accounts of desertification pin the blame on the actions of ethnic minority herders and how few foreign scientists question such accounts.

21. The concept of *chuantong zhishi* was as much a creative invention as the concept of indigenous knowledge, for in China the idea of traditional had little appeal during the Mao era, when many buildings and objects associated with Taoism, Confucianism, and Buddhism were destroyed. Only in the 1990s did the notion of Chinese traditional culture really gain a strong positive force, and this was almost always associated with aspects of elite culture and power, especially emblems of Chinese identity that circulated internationally, like calligraphy, the Beijing Opera, and the Great Wall.

22. Yu made his argument, coincidentally, just as a scandal emerged in the United States (Colchester 1994) detailing the often forgotten history that Yellowstone

National Park was not always unpeopled and wild but created by the forced eviction of indigenous peoples. Such knowledge was well-known to early park visitors, who traveled under guard of the U.S. Army, still nervous about the remaining Lakota Indians. Over time this past was largely forgotten and became a scandal only nearly a hundred years after the eviction because of significant shifts in public sentiments, including white guilt for the past and growing acknowledgment of American colonialism.

23. This notion of activism and civil society parallels scholars' assumptions that rural residents resisted the state's intrusions and craved autonomy, as discussed in chapter 3.

24. The Constitution of the PRC (adopted on December 4, 1982) declares, "The state maintains public order and suppresses treasonable and other counter-revolutionary activities; it penalizes actions that endanger public security and disrupt the socialist economy and other criminal activities, and punishes and reforms criminals." The reference to "counter-revolutionary" was removed in 1999, as China reduced Maoist language, http://www.npc.gov.cn/englishnpc/Constitution/node_2827.htm (accessed September 6, 2012).

25. One exception was during the Cultural Revolution, when the term was frequently used to target others, even by children against their own parents.

26. These estimates vary widely, from 90,000 to 180,000, as does the definition of an "incident," which is not always a protest. Using some definitions, a single protestor can constitute an incident; in other definitions, a minimum of a hundred people are required. Regardless, this is a truly remarkable figure, especially when released by a government that often works quite hard to present an image of domestic harmony.

CHAPTER FIVE

1. Human deaths are described in a 2004 report to the UNESCO Secretariat. Of course, in many places around the world, elephants are much more likely to be the victims of lethal attacks by humans than the other way around. In this case, I was both surprised to see the high rate of elephant aggression and curious to explore villagers' claims that elephant aggression had markedly increased.

2. See http://www.china.org.cn/china/2011-11/07/content_23842199.htm (accessed January 21, 2012). This figure concerns grain that elephants destroyed but did not necessarily consume. I saw several cases where a single elephant destroyed virtually all the grain in a three-*mu* plot (one hectare is fifteen mu). The figures imply that every year, every elephant out of two hundred destroyed an average of 25,000 pounds of grain. Based on estimated paddy yields of 5,000 pounds per hectare (Xu 2004), that would mean each elephant destroying five hectares of grain in paddy per year. Approximately one million people live in Xishuangbanna, so this is about 100 pounds of grain per person. It seems like a high figure but possible.

3. Elephants may stimulate the thickest webs of affective connections of any nonhuman species in Yunnan, just as giant pandas (which live in Sichuan Province) would be the most prominent animal for China as a whole. From an anthropocentric perspective, we should also consider that some of the most important species might not be charismatic megafauna; they may be quite small, like the malarial mosquito.

4. Many transnational engagements with wild animals are driven not solely by nature conservation efforts but by a diverse range of actors with overlapping and conflicting aims. For instance, WWF staff privately dismiss some animal welfare advocates as "humaniacs," punning on *humane* and *maniacs*, because they privilege animal welfare for favored domestic species above ecosystem protection or wild species. Thus these groups, ostensibly all "environmental," strive to create different social worlds. In China animal-focused groups now include animal welfare organizations, zoos, hunters, consumers of Chinese medicine and animal-derived products, and public health officials concerned with animals as disease vectors.

5. The story of scallops in France is more complex, as some were successfully domesticated before disease wiped them out (Alban and Boncoeur 2008).

6. Books like Sidney Mintz's *Sweetness and Power: The Place of Sugar in Modern History* (1985) explore colonialism's impact on European social and culinary life. However, Prestholdt's close examination of how Africans interpret, use, and desire particular European goods, and how these desires effect change in Europe, has few precedents.

7. For the argument that these acts represent "animal resistance," see Jason Hribal's book, *Fear of the Animal Planet: The Hidden History of Animal Resistance* (2011). While I agree with Hribal that nonhumans possess forms of agency, I am leery of classifying certain acts as resistance, especially as we have little ability to understand nonhumans' perceptions of the world and motivations for acting in particular ways.

8. Bennett (2004: 347) and others study a range of actants, exploring the "vitality, willfullness, and recalcitrance possessed by nonhuman entities and forces."

9. I draw on scholarship by environmental historians who are trying to understand nonhumans as historical actors (Cronon 1983; Crosby 2004). For an exemplary study of how animals shape human lives and histories, see Virginia Anderson's *Creatures of Empire: How Domestic Animals Transformed Early America* (2006).

10. This plane was likely left over from World War II, showing more layers of Yunnan's transnational history, as it was favored by the Flying Tigers.

11. There were plans to build an observation center for guar, a wild cattle ancestor that is also an endangered species. The viewing spot has not materialized, however, as people seem to have far less desire to see guar than elephants.

12. There was some mention of China's elephants in surveys from the 1950s, but no systematic studies and few data points (Shou et al. 1959). It is unclear if WWF staff ever came across these older reports, which would be difficult to find, as the primary author died during the Cultural Revolution. In the 1990s I still found it

difficult to track down older papers, as library collections were often incomplete and many articles were published in obscure journals with small print runs. Obtaining these articles often required contacting the authors and visiting their homes.

13. Before Santiapillai, the only other international study of Asian elephants was by Cambridge biologist Robert Oliver in the 1970s. Oliver mentioned China's surveys from the 1950s but did not know if elephants had become extinct since then.

14. In 1999 Santiapillai believed the number of elephants to be around three hundred, revising his earlier estimate. He stated that this lower number was not because he actually believed that elephants had died off but because surveys are quite difficult to carry out in the thick forest, and the area is vast. In 2003 a report of which Santiapillai was a coauthor, by an NGO called Elephant Family, estimated that China had between 200 and 250 elephants (Jepson and Canney 2003).

15. A similar attitude was described in chapter 4, when Beijing photojournalists in Banna described local groups' actions as wasteful and destructive (e.g., Tang 1987).

16. Starting in 1987 government officials declared that killing giant pandas could be a capital crime. In 1997 Amnesty International revealed that the government executed three men involved in the panda trade, right before China downgraded killing endangered species from a capital offense. In Africa, however, shoot-to-kill orders for poachers of elephants and other big game have resulted in many deaths over the years.

17. The poaching was likely for tusks, although it was carried out at great risk and for relatively little benefit, as Asian elephants have much smaller tusks, on average, than African elephants. Recent studies, in fact, suggest that over the generations, elephants are evolving to have smaller tusks due to long-term hunting pressure on those with larger tusks. According to villagers, the poachers may have also been motivated by a sense of vigilantism. Over time, the executed men have become seen as martyrs, as those who "fought back" against elephants when state officials did not seem to show much concern for rural livelihoods.

18. It was often said that male elephants were more likely than females to be aggressive to humans and thus labeled as rogues. This is likely related to the males' musth cycle, when they are interested in mating and discharge a thick fluid from behind the ears. During musth, the animal can be quite violent. I never heard anyone in China talk about musth, but there is much discussion elsewhere in Asia, especially where elephants are trained.

19. In 1983 a federal law outlawed hunting "rare animals," which included elephants. In 1991 Yunnan Province banned hunting, but researchers found that villagers and officials in upland areas had little knowledge of or respect for such laws (Harris and Ma 1997).

20. Bandits were especially interested in the wealth from opium, a key part of the regional economy before the 1950s. According to a nature reserve guard, the arrival of the People's Liberation Army eventually eliminated opium and gangs of bandits, but villagers maintained their guns, in part due to rumors that Guomindang soldiers in Thailand, still loyal to Chiang Kai Shek and supported by the CIA,

plotted to invade Yunnan. As well, when Chinese troops invaded Vietnam in 1979, supposedly in self-defense, some local officials gave guns to villagers.

21. I was told that the earlier law, before the confiscation, required guns to be legally registered, but it resulted in almost no registrations. In fact few people seemed aware of it, and I never heard of a villager registering a gun.

22. Elephants are remarkably tough. After an elephant in Hawai'i escaped from a city parade, police shot it eighty-six times before it finally stopped running (Bernardo 2004). Police use special bullets and far more powerful guns than the muzzleloaders in Xishuangbanna.

23. Interestingly, there is support for their suspicion, as recent newspaper articles suggest that "more than 30" people have been killed by elephants in Yunnan, http://en.kunming.cn/index/content/2011-11/25/content_2751505_5.htm (accessed June 3, 2012). An in-depth report, which I describe below, however, shows that at least sixty-three people have been killed (Secretariat 2004).

24. Like much of China, nearly every rural family relies on the grain harvest for their daily food. Until 2005 there was 1 million yuan available for the prefecture, an average of 1 yuan per person (at approximately 7 or 8 yuan to the USD, about 15 cents). Farmers received about 1 yuan for every 2 kilograms of grain destroyed. In 2005 the federal government allocated an additional 4 million yuan for compensation, particularly for damage caused by elephants. When an elephant killed a person, the family might be allocated 12,000 yuan, but most funerals cost more (Anonymous 2006).

25. In 2003 the park was purchased by a firm from Zhejiang Province, which then contracted out particular features, such as selling management rights of the aerial tram ride to a company from Shandong Province. Thus the park is no longer under the auspices of the Nature Reserve Bureau or local ownership.

26. In comparison, in southern India thirty to fifty people are killed by elephants every year, but the overall number of elephants is much greater (approximately seven thousand). Thus in southern India seven thousand elephants kill an average of forty people, making the ratio approximately 1:200 of elephants to people killed. In Xishuangbanna two hundred elephants kill an average of ten people a year, making the ratio 1:20; thus China's elephants kill ten times more people, likely making this particular population among the deadliest group of animals of any species on the planet.

27. See http://news.xinhuanet.com/english2010/china/2011-01/04/c_13676553.htm (accessed January 31, 2012).

28. Unlike giant pandas, however, China does not monopolize Asian elephants, and so it is less likely that this will emerge as a major international center. Like the Panda Center, the elephant center received criticism from some foreign wildlife biologists, such as George Schaller (1993) and Richard Harris (2008), who charge China with overly prioritizing captive breeding programs and underfunding habitat preservation.

29. See http://www.eleaid.com/index.php?page=elephantsinchina (accessed October 12, 2011).

GLOSSARY

bijiao fada	比较发达
cha laoban	茶老板
chi	吃
chuangtong	传统
chuangtong zhishi	传统知识
chui niu	吹牛
congming	聪明
danzi da	胆子大
dingzi	钉了
Dong Feng	东风
fan geming	反革命
fen lie	分裂
feng shui	风水
fengjian yishi	封建意识
gaige kaifang	改革开放
gan die	干爹
geming qunzhong	革命群众
guanxi	关系
guanxi xue	关系学
heli kaifa	合理开发
hetong dingxiale	合同订下了
huang niu	黄牛
huanjing feng	环境风
Huaxia Renwen Dili	华夏人文地理

huodong yuan	活动员
huodong zhe	活动者
jiefang	解放
jijifenzi	积极分子
jizi	麂子
kai huangdi	开荒地
kao	拷
kuaiji	会计
la guanxi	拉关系
lao chou jiu	老臭九
laobaixing	老百姓
laoban	老板
Lolo	*originally* 猓猓, then 倮倮 then 罗罗
Long Shan	龙山
long yan	龙睛
luohuo	落后
maodun	矛盾
Meiguoren	美国人
meiyou wenhua	没有文化
mianzi	面子
nongmin	农民
rang nongmin	让农民
re	热
Renmin Shiwu Heszoshe	人民食物合作社
san da fa	三大伐
sha ren	砂仁
shaoshu minzu	少数民族
shou fu tuolaji	手扶拖拉机
si hai jiu zhou	四海九州
sixiang gongzuo	思想工作
su ku	诉苦
tiaopi	调皮
tongzhi	同志
touxiang	偷相
tufei	土匪

tuzhu ren	土著人
waiguoren	外国人
wailai sixiang	外来思想
wangguo	王国
wenhua zhan	文化站
wopeng	窝棚
Xi Feng	西风
xiangmu	项目
Xiao Long	小龙
ye niu	野牛
yi bai bing	一百病
Yinwei tamen de fengjian mixin	因为他们的封建迷信
yuan	元
yundong	运动
zangde	脏的
zhongguo guoqing	中国国情
zhu shu	竹鼠

BIBLIOGRAPHY

Abu-Lughod, Janet. 1989. *Before European Hegemony: The World System 1250–1350*. New York: Oxford University Press.

Abu-Lughod, Lila. 1990. "The Romance of Resistance: Tracing Transformations of Power through Bedouin Women." *American Ethnologist* 17 (July): 41–55.

Adams, J. S., and T. O. McShane. 1992. *The Myth of Wild Africa: Conservation without Illusion*. Berkeley: University of California Press.

Adams, V., and S. Pigg, eds. 2006. *Sex in Development: Science, Sexuality and Morality in Global Perspective*. Durham, NC: Duke University Press.

Aird, J. S. 1990. *Slaughter of the Innocents: Coercive Birth Control in China*. Washington, DC: American Enterprise Institute.

Alban, F., and J. Boncoeur. 2008. *Sea-Ranching in the Bay of Brest (France): Technical Change and Institutional Adaptation of a Scallop Fishery*. FAO Fisheries Technical Paper.

Anagnost, A. 1997. *National Past-Times*. Durham, NC: Duke University Press.

Anderson, D., and R. H. Grove, eds. 1989. *Conservation in Africa: Peoples, Policies and Practice*. Cambridge, UK: Cambridge University Press.

Anderson, D. M., J. Salick, R. K. Moseley, and O. Xiaokun. 2005. "Conserving the Sacred Medicine Mountains: A Vegetation Analysis of Tibetan Sacred Sites in Northwest Yunnan." *Biodiversity and Conservation* 14 (13): 3065–91.

Anderson, V. D. J. 2006. *Creatures of Empire: How Domestic Animals Transformed Early America*. New York: Oxford University Press.

Anonymous. 1979. "Protecting the Forests." *Beijing Review* 22 (9): 4.

Anonymous. 1991. *Conservation in Xishuangbanna, China*. Geneva: World Wildlife Fund.

Anonymous. 2001 云南省西双版纳州收缴销毁六万余支枪支 (Yunnan sheng xishuangbanna zhou shoujiao xiaohui liu wan yu zhi qiangzhi) Yunnan Province Xishuangbanna Prefecture More than 60,000 Seized Firearms were Destroyed, http://news.sina.com.cn/c/213850.html (accessed February 10, 2013).

Anonymous. 2006. "Elephants: Revered Icon or Dangerous Threat?." *China Daily*, June 7, http://www.china.org.cn/english/environment/170661.htm (accessed February 29, 2012).

Apel, Ulrich. 2000. *Traditional Village Forest Management: The Village Forest of Moxie, Southwest China*. Eschborn: German Agency for Technical Cooperation.

Appadurai, Arjun. 1996. *Modernity at Large: Cultural Dimensions of Globalization*. Minneapolis: University of Minnesota Press.

Arkush, R. David. 1981. *Fei Xiaotong and Sociology in Revolutionary China*. Cambridge, MA: Harvard University Press.

Ashcroft, B. 2001. *Post-colonial Transformation*. London: Routledge.

Askew, Kelly M. 2002. *Performing the Nation: Swahili Music and Cultural Politics in Tanzania*. Chicago: University of Chicago Press.

Austin, C. J. 2006. *Up Against the Wall: Violence in the Making and Unmaking of the Black Panther Party*. Little Rock: University of Arkansas Press.

Bamford, Dave. 1988. "WWF Project 3194: China, Xishuangbanna Reserves, Nature Tourism Consultancy." Unpublished report for WWF-China (Hong Kong).

Baranovitch, Nimrod. 2001. "Between Alterity and Identity: New Voices of Minority People in China." *Modern China* 27 (3): 359–401.

Baranovitch, Nimrod. 2010. "Others No More: The Changing Representation of Non-Han Peoples in Chinese History Textbooks, 1951–2003." *The Journal of Asian Studies* 69 (1): 85–122.

Bargh, M. 2007. *Resistance: An Indigenous Response to Neoliberalism*. Wellington, NZ: Huia Publishers.

Barnes, R. H., Andrew Gray, and Benedict Kingsbury, eds. 1995. *Indigenous Peoples of Asia*. Ann Arbor, MI: Association for Asian Studies.

Baviskar, A. 2005. "Adivasi Encounters with Hindu Nationalism in MP." *Economic and Political Weekly* 40 (48): 5105–13.

Bennett, Jane. 2004. "The Force of Things Steps toward an Ecology of Matter." *Political Theory* 32 (3): 347–372.

Bennett, J. 2009. *Vibrant Matter: A Political Ecology of Things*. Durham, NC: Duke University Press.

Bentley, Jack. 1990. "Agroforestry Handbook for Xishuangbanna." Unpublished Report for WWF-China (Hong Kong).

Bentley, Jack. 1993. *Xishuangbanna Agroforestry Development Project: A Strategy and Procedure for Implementation*. Unpublished Report for WWF-China (Hong Kong).

Berlant, L., and M. Warner. 1998. "Sex in Public." *Critical Inquiry* 24 (2): 547–66.

Bernardo, Rosemarie. 2004. "Shots Killing Elephant Echo across a Decade." *Star Bulletin*, Honolulu, HI. August 16.

Bernstein, H. 1990. "Taking the Part of Peasants." In *The Food Question: Profit versus People?*, 69–79. New York: Monthly Review Press.

Biehl, J. 2005. *Vita: Life in a Zone of Social Abandonment*. Berkeley: University of California Press.

Blaikie, Piers, and Harold Brookfield. 1987. *Land Degradation and Society*. New York: Methuen.

Blaikie, Piers, and Joshua Muldavin. 2004. "Upstream, Downstream, China, India: The Politics of Environment in the Himalayan Region." *Annals of the Association of American Geographers* 94 (3): 520–48.

Blaut, J. M., ed. 1992. *1492: The Debate on Colonialism, Eurocentrism, and History*. Trenton, NJ: Africa World Press.

Blum, S. D. 2000. *Portraits of "Primitives": Ordering Human Kinds in the Chinese Nation*. Lanham, MD: Rowman & Littlefield.

Boellstorff, T. 2005. *The Gay Archipelago: Sexuality and Nation in Indonesia*. Princeton, NJ: Princeton University Press.

Boland, A. 2000. "Feeding Fears: Competing Discourses of Interdependency, Sovereignty, and China's Food Security." *Political Geography* 19 (1): 55–76.

Bolin, A., and J. Granskog. 2003. *Athletic Intruders: Ethnographic Research on Women, Culture, and Exercise*. Albany: State University of New York Press.

Bonner, R. 1993. *At the Hand of Man: Peril and Hope for Africa's Wildlife*. New York: Knopf.

Borchert, Thomas. 2008. "Worry for the Dai Nation: Sipsongpannā, Chinese Modernity, and the Problems of Buddhist Modernism." *The Journal of Asian Studies* 67 (1): 107–142.

Borchert, Thomas. 2010. "The Abbot's New House: Thinking about How Religion Works among Buddhists and Ethnic Minorities in Southwest China." *Journal of Church and State* 52 (1): 112–137.

Bottelier, Pieter. 2001. "Was World Bank Support for the Qinghai Anti-Poverty Project in China Ill-Considered?" *Harvard Asia Quarterly* 5 (1): 47–55.

Boxer, B. 1981. "Environmental Science and Policy in China." *Environment: Science and Policy for Sustainable Development* 23 (5): 14–37.

Brady, Anne-Marie. 2003. *Making the Foreign Serve China: Managing Foreigners in the People's Republic*. Lanham, MD: Rowman & Littlefield.

Bray, Francesca. 1986. *The Rice Economies: Technology and Development in Asian Societies*. New York: Basil Blackwell.

Brockington, D., and J. Igoe. 2006. "Eviction for Conservation: A Global Overview." *Conservation and Society* 4 (3): 424–70.

Brosius, J. Peter. 1997. "Prior transcripts, divergent paths: Resistance and acquiescence to logging in Sarawak, East Malaysia." *Comparative Studies in Society and History* 39 (3): 468–510.

Brosius, J. P. 2004. "Indigenous Peoples and Protected Areas at the World Parks Congress." *Conservation Biology* 18 (3): 609–12.

Brysk, Alison. 2000. *From Tribal Village to Global Village: Indian Rights and International Relations in Latin America*. Stanford: Stanford University Press.

Bulag, U. E. 2012. "Good Han, Bad Han: The Moral Parameters of Ethnopolitics in China." In *Critical Han Studies*, ed. Thomas Mullaney, James Leibold, Stéphane Gros, and Eric Vanden Bussche, 92–112. Berkeley: University of California Press.

Callon, M. 1986. "Some Elements of a Sociology of Translation: Domestication of the Scallops and the Fisherman of St. Brieuc Bay." In *Power, Action and Belief: A New Sociology of Knowledge*, ed. J. Law, 196–223. London: Routledge.

Caro, T. M., and G. O'Doherty. 1999. "On the Use of Surrogate Species in Conservation Biology." *Conservation Biology* 13 (4): 805–14.

Cepek, Michael L. 2011. "Foucault in the Forest: Questioning Environmentality in Amazonia." *American Ethnologist* 38 (3): 501–15.

Chabot, S. 2004. "Framing, Transnational Diffusion, and African-American Intellectuals in the Land of Gandhi." *International Review of Social History* 49 (12): 19–40.

Chabot, S., and J. W. Duyvendak. 2002. "Globalization and Transnational Diffusion between Social Movements: Reconceptualizing the Dissemination of the Gandhian Repertoire and the 'Coming Out' Routine." *Theory and Society* 31 (6): 697–740.

Chadwick, D. H. 1994. *The Fate of the Elephant*. San Francisco: Sierra Club Books.

Chakrabarty, D. 2008. *Provincializing Europe: Postcolonial Thought and Historical Difference*. Princeton, NJ: Princeton University Press.

Chambers, Robert, and M. Leach. 1989. "Trees as Savings and Security for the Rural Poor." *World Development* 17 (3): 329–42.

Chao, Emily. 1996. "Hegemony, Agency, and Re-presenting the Past: The Invention of the Dongba Culture among the Naxi of Southwest China." In *Negotiating Ethnicities in China and Taiwan*, ed. Melissa Brown, 208–39. Berkeley: Institute of East Asian Studies, University of California, Berkeley.

Chapin, Mac. 2004. "A Challenge to Conservationists." *World Watch*, November/December: 17–31.

Chen, Sanyang, Shengji Pei, and Jianchu Xu. 1992. "Indigenous Management of the Rattan Resources in the Forest Lands of Mountain Environment: The Hani Practice in the Mengsong Area of Yunnan, China." *Ethnobotany* 4: 412–22.

Cheng, Xiao-fang, Yu-ming Yang, Ying Huang, and Juan Wang. 2008. "On the Traditional Utilization and Management of Forest Resources of Minority Nationalities Live in Communities Around Lancang River Nature Reserve." *Journal of Beijing Forestry University (Social Sciences)* 7 (1): 8–12.

Cheo, R. 2000. *An Evaluation of the Impact of Rubber Trees in China on the Rural Economy with Specific Focus on Xishuangbanna, Yunnan and Hainan Island*. Working Paper. National University of Singapore.

Cheung, Pui Shan Catherine, and John MacKinnon. 1991. "Xishuangbanna Nature Conservation: Exhibition Report." Unpublished Report for WWF-China (Hong Kong).

Choy, T. K. 2011. *Ecologies of Comparison: An Ethnography of Endangerment in Hong Kong*. Durham, NC: Duke University Press.

Clarke, W. C. 1976. "Maintenance of Agriculture and Human Habitats within the Tropical Forest Ecosystem." *Human Ecology* 4 (3): 247–59.

Claudio, L. 2007. "Standing on Principle: The Global Push for Environmental Justice." *Environmental Health Perspectives* 115 (10): A500–A503.

Clifford, James. 1988. *The Predicament of Culture: Twentieth Century Ethnography, Literature, and Art.* Cambridge, MA: Harvard University Press.

Clifford, J. 2001. "Indigenous Articulations." *The Contemporary Pacific* 13 (2): 468–90.

Cobb, D. M. 2008. *Native Activism in Cold War America: The Struggle for Sovereignty.* Lawrence: University Press of Kansas.

Coggins, C. 2002. *The Tiger and the Pangolin: Nature, Culture, and Conservation in China.* Honolulu: University of Hawaii Press.

Cohen, M. L. 1993. "Cultural and Political Inventions in Modern China: The Case of the Chinese Peasant." *Daedalus* 122: 151–70.

Colchester, Marcus. 1994. *Salvaging Nature: Indigenous Peoples, Protected Areas, and Biodiversity Conservation.* Geneva: UN Research Institute for Social Development.

Colchester, Marcus, Fergus MacKay, Tom Griffiths, and John Nelson. 2001. *A Survey of Indigenous Land Tenure.* Unpublished Report. Rome: Forest Peoples Programme and FAO.

Comaroff, J. 1985. *Body of Power, Spirit of Resistance: The Culture and History of a South African People.* Chicago: University of Chicago Press.

Conklin, Beth A., and Laura R. Graham. 1995. "The Shifting Middle Ground: Amazonian Indians and Eco-politics." *American Anthropologist* 97 (4): 1–17.

Cooper, F. 2005. *Colonialism in Question.* Berkeley: University of California Press.

Cooper, Frederick, and Ann Stoler, eds. 1997. *Tensions of Empire: Colonial Cultures in a Bourgeois World.* Berkeley: University of California Press.

Corbridge, S., G. Williams, M. Srivastava, and R Veron. 2004. *Seeing the State: Governance and Governmentality in Rural India.* Cambridge, UK: Cambridge University Press.

Croll, Elisabeth, and David Parkin, eds. 1992. *Bush Base: Forest Farm. Culture, Environment, and Development.* London: Routledge.

Cronon, William. 1983. *Changes in the Land: Indians, Colonists, and the Ecology of New England.* New York: Hill and Wang.

Crosby, Alfred W. 2004. *Ecological Imperialism: The Biological Expansion of Europe, 900–1900.* 2nd ed. Cambridge, UK: Cambridge University Press.

Dave, Naisargi N. 2011. "Indian and Lesbian and What Came Next: Affect, Commensuration, and Queer Emergences." *American Ethnologist* 38 (4): 650–65.

Davis, K. 2007. *The Making of Our Bodies, Ourselves: How Feminism Travels across Borders.* Durham, NC: Duke University Press.

Davoll, J. 1988. "Population Growth and Conservation Organizations, with Particular Reference to the International Union for Conservation of Nature and Natural Resources (IUCN)." *Population & Environment* 10 (2): 107–14.

Dikötter, Frank. 1992. *The Discourse of Race in Modern China.* Stanford: Stanford University Press.

Dikötter, F. 2008. *The Age of Openness: China before Mao.* Berkeley: University of California Press.

Dirlik, Arif. 1978. *Revolution and History: The Origins of Marxist Historiography in China, 1919–1937.* Berkeley: University of California Press.

Dirlik, Arif. 1996. "Reversals, Ironies, Hegemonies: Notes on the Contemporary Historiography of Modern China." *Modern China* 22 (3): 243–84.

Douglas, M. 1970. *Natural Symbols*. London: Routledge.

Eckwall, E. 1955. "Slash and Burn Cultivation: A Contribution to the Anthropological Terminology." *Man* 55: 135–36.

Economy, Elizabeth C. 2004. *The River Runs Black: The Environmental Challenge to China's Future*. Ithaca, NY: Cornell University Press, Council of Foreign Relations.

Edmonds, Richard Louis. 2012. *Patterns of China's Lost Harmony: A Survey of the Country's Environmental Degradation and Protection*. London: Routledge.

Ehrlich, P. R. 1968. *The Population Bomb*. New York: Ballantine Books.

Ehrlich, Paul R., and Anne H. Ehrlich. 2009. "The Population Bomb Revisited." *Electronic Journal of Sustainable Development* 1 (3): 63–71.

Elvin, M. 2004. *The Retreat of the Elephants: An Environmental History of China*. New Haven, CT: Yale University Press.

Erazo, Juliet S. Forthcoming. *Governing Indigenous Territories: Enacting Sovereignty in the Ecuadorian Amazon*. Durham, NC: Duke University Press.

Esposito, B. J., and T. T. Lie. 1971. *The Cultural Revolution and Science Policy and Development in Mainland China*. Brussels: Centre d'étude du Sud-Est asiatique et de l'Extrême-Orient, Institut de sociologie de l'Université libre de Bruxelles.

Faier, L. 2009. *Intimate Encounters: Filipina Women and the Remaking of Rural Japan*. Berkeley: University of California Press.

Fan, Fa-ti. 2004. *British Naturalists in Qing China: Science, Empire, and Cultural Encounter*. Cambridge, MA: Harvard University Press.

Fan, F. 2012. "'Collective Monitoring, Collective Defense': Science, Earthquakes, and Politics in Communist China." *Science in Context* 25 (1): 127–54.

Featherstone, Mike. 2006. "Genealogies of the Global." *Theory, Culture & Society* 23 (2–3): 387–92.

Fenelon, J. V., and T. D. Hall. 2008. "Revitalization and Indigenous Resistance to Globalization and Neoliberalism." *American Behavioral Scientist* 51 (12): 1867–901.

Ferguson, J. 2006. *Global Shadows: Africa in the Neoliberal World Order*. Durham, NC: Duke University Press.

Ferguson, James. 1994. *The Anti-Politics Machine: "Development," Depoliticization, and Bureaucratic Power in Lesotho*. Minneapolis: University of Minnesota Press.

Flusty, S. 2004. *De-Coca-Colonization: Making the Globe from the Inside Out*. New York: Routledge.

Foster, R. J. 2008. *Coca-Globalization: Following Soft Drinks from New York to New Guinea*. New York: Palgrave Macmillan.

Foucault, Michel. 1977. *Discipline and Punish: The Birth of the Prison*. New York: Pantheon.

Friedman, J. 1999. "Indigenous Struggles and the Discreet Charm of the Bourgeoisie." *Australian Journal of Anthropology* 10 (1): 1–14.

Friedman, T. L. 2006. *The World Is Flat: The Globalized World in the Twenty-first Century*. London: Penguin.

Fung, K. I. 1972. "China's Grain Trade: Explanation and Prospects." *Canadian Geographer/Le Géographe Canadien* 16 (1): 15–28.

Gadgil, Madhav, and V. D. Vartak. 1976. "Sacred Groves of the Western Ghats of India." *Economic Botany* 30 (1): 152–60.

Gansu Province Health Department. 1999. "China: Rural Health Project Indigenous Peoples Planning Framework (IPPF)," http://www.gsws.gov.cn/html/3/28/5591.htm (accessed October 12, 2012).

Gao, Hongzhi. 1998. "Towards Sustainable Communities: Environmental and Resource Management in Lijiang China." M.A. thesis, Simon Fraser University.

Gao, Lishi. 1999. *On the Dai's Traditional Irrigation System and Environmental Protection in Xishuangbanna*. Kunming, Yunnan: Yunnan Minzu Chubanshe.

García, Maria Elena. 2005. *Making Indigenous Citizens: Identities, Education, and Multicultural Development in Peru*. Stanford: Stanford University Press.

Garland, E. 2008. "The Elephant in the Room: Confronting the Colonial Character of Wildlife Conservation in Africa." *African Studies Review* 51 (3): 51–74.

Geertz, Clifford. 1963. *Agricultural Involution: The Processes of Ecological Change in Indonesia*. Berkeley: University of California Press.

Gibson-Graham, J. K. 1996. *The End of Capitalism (As We Knew It)*. London: Blackwell.

Gibson-Graham, J. K. 2006. *A Postcapitalist Politics*. Minneapolis: University of Minnesota Press.

Giersch, C. P. 2006. *Asian Borderlands: The Transformation of Qing China's Yunnan Frontier*. Cambridge, MA: Harvard University Press.

Givan, R. K., K. M. Roberts, and S. A. Soule, eds. 2010. *The Diffusion of Social Movements: Actors, Mechanisms, and Political Effects*. Cambridge, UK: Cambridge University Press.

Gladney, Dru. 1994. "Representing Nationality in China: Refiguring Majority/Minority Identities." *Journal of Asian Studies* 53 (1): 92–123.

Goldman, M. 2001. "The Birth of a Discipline: Producing Authoritative Green Knowledge, World Bank–Style." *Ethnography* 2 (2): 191–218.

Graham, L. 2002. "How Should an Indian Speak? Brazilian Indians and the Symbolic Politics of Language Choice in the International Public Sphere." *Indigenous Movements, Self-Representation, and the State in Latin America*, ed. K. Warren and J. Jackson, 181–228. Austin: University of Texas Press.

Grant, Bruce. 1995. *In the Soviet House of Culture*. Princeton, NJ: Princeton University Press.

Greenhalgh, S. 2003. "Planned Births, Unplanned Persons: Population in the Making of Chinese Modernity." *American Ethnologist* 30 (2): 196–215.

Greenhalgh, S. 2008. *Just One Child: Science and Policy in Deng's China*. Berkeley: University of California Press.

Greenhalgh, S. 2010. *Cultivating Global Citizens: Population in the Rise of China*. Cambridge, MA: Harvard University Press.

Grove, Richard. 1995. *Green Imperialism: Colonial Expansion, Tropical Island Edens and the Origins of Environmentalism, 1600–1860*. Cambridge, UK: Cambridge University Press.

Guha, Ranajit. 1997. *Dominance without Hegemony: History and Power in Colonial India*. Cambridge, MA: Harvard University Press.

Guha, Ramachandra. 1990. *Unquiet Woods: Ecological Change and Peasant Resistance in the Indian Himalaya*. Berkeley: University of California Press.

Guha, Ranajit, and Gayatri Chakravorty Spivak, eds. 1988. *Selected Subaltern Studies*. Oxford: Oxford University Press.

Guillain, R. 1957. *The Blue Ants*. London: Secker and Warburg.

Guo, Jiaji. 1998. *Rice-Farming Culture of the Dai People in Xishuangbanna (Bilingual Edition)*. Kunming, Yunnan: Yunnan Daxue Chubanshe.

Gupta, Akhil. 1995. "Blurred Boundaries: The Discourse of Corruption, the Culture of Politics, and the Imagined State." *American Ethnologist* 22 (2): 375–402.

Gupta, A., and A. Sharma. 2006. "Globalization and Postcolonial States." *Current Anthropology* 47 (2): 277–307.

Guthman, Julie. 1997. "Representing Crisis: The Theory of Himalayan Environmental Degradation and the Project of Development in Post-Rana Nepal." *Development and Change* 28 (1): 45–69.

Hamilton, J. A. 2008. *Indigeneity in the Courtroom: Law, Culture, and the Production of Difference in North American Courts*. New York: Routledge.

Han, D. 2008. *The Unknown Cultural Revolution: Life and Change in a Chinese Village*. New York: Monthly Review Press.

Hanisch, C. 2006. "Updated Introduction to 'The Personal Is Political,'" http://www.carolhanisch.org/CHwritings/PIP.html (accessed February 3, 2013).

Hanisch, Carol. 1970. "The Personal Is Political." In *Notes from the Second Year: Women's Liberation*, ed. Shulamith Firestone and Anne Koedt, 204–5. New York: Radical Feminism.

Hannerz, U. 2002. *Flows, Boundaries and Hybrids: Keywords in Transnational Anthropology*. Oxford: University of Oxford, Transnational Communities Programme.

Hansen, Mette Halskov. 1999. *Lessons in Being Chinese*. Seattle: University of Washington Press.

Hansen, Mette Halskov. 2005. *Frontier People: Han Settlers in Minority Areas of China*. Vancouver: University of British Columbia Press.

Haraway, Donna. 1989. *Primate Visions: Gender, Race, and Nature in the World of Modern Science*. New York: Routledge.

Haraway, Donna J. 2008. *When Species Meet*. Minneapolis: University of Minnesota Press.

Harper, Janice. 2002. *Endangered Species: Health, Illness, and Death among Madagascar's People of the Forest*. Durham, NC: Carolina Academic Press.

Harrell, Steven, ed. 1995. *Cultural Encounters on China's Ethnic Frontiers*. Seattle: University of Washington Press.

Harris, R. B. 2008. *Wildlife Conservation in China: Preserving the Habitat of China's Wild West*. Armonk, NY: M. E. Sharpe.

Harris, Richard B., and Shilai Ma. 1997. "Initiating a Hunting Ethic in Lisu Villages, Western Yunnan, China." *Mountain Research and Development* 17 (2): 171–76.

Hart, G. P. 2002. *Disabling Globalization: Places of Power in Post-Apartheid South Africa*. Berkeley: University of California Press.

Harvey, David. 1989. *The Conditions of Postmodernity: An Enquiry into the Origins of Cultural Change*. Oxford: Blackwell.

Harvey, D. 2000. *Spaces of Hope*. New York: Taylor & Francis.

Hathaway, Michael J. 2010a. "The Emergence of Indigeneity: Public Intellectuals and an Indigenous Space in Southwest China." *Cultural Anthropology* 25 (2): 301–33.

Hathaway, Michael J. 2010b. "Global Environmental Encounters in Southwest China: Fleeting Intersections and 'Transnational Work.'" *Journal of Asian Studies* 69 (2): 427–51.

Hathaway, Michael J. 2011. "Global Environmentalism and the Emergence of Indigeneity." In *The Anthropology of Extinction: Essays on Culture and Species Death*, ed. G. M. Sodikoff, 103–26. Bloomington: Indiana University Press.

Haugerud, Angelique. 2005. "Globalization and Thomas Friedman." In *Why America's Top Pundits Are Wrong: Anthropologists Talk Back*, ed. C. L. Besteman and H. Gusterson, 102–20. Berkeley: University of California Press.

Hay-Edie, Terence. 2004. "International Animation: UNESCO, Biodiversity and Sacred Sites." In *Development and Local Knowledge: New Approaches to Issues in Natural Resources Management, Conservation and Agriculture*, ed. A. Bicker, P. Sillitoe, and J. Pottier, 119–34. London: Routledge.

Hayden, Cori. 2003. *When Nature Goes Public: The Making and Unmaking of Bioprospecting in Mexico*. Berkeley: University of California Press.

Hays, Samuel P. 1987. *Beauty, Health, and Permanence: Environmental Politics in the United States, 1955–1985*. Cambridge, UK: Cambridge University Press.

Hays, Samuel P. 1999. *Conservation and the Gospel of Efficiency: The Progressive Conservation Movement, 1890–1920*. Pittsburgh: University of Pittsburgh Press.

He, Di. 1994. "The Most Respected Enemy: Mao Zedong's Perception of the United States." *China Quarterly* 137 (March): 144–58.

He, S. H., and H. J. Li. 2011. "Hymenochaete rhododendricola and H. quercicola spp. nov. (Basidiomycota, Hymenochaetales) from Tibet, Southwestern China." *Nordic Journal of Botany* 29 (4): 484–87.

He, Xing-liang. 2004. "On the Traditional Culture of China's Ethnic Minorities and Ecological Protection." *Journal of Yunnan University for Nationalities (Social Sciences)* 21 (1): 48–56.

Hecht, Susanna, and Alexander Cockburn. 1990. *The Fate of the Forest: Developers, Destroyers and Defenders of the Amazon*. New York: HarperCollins.

Hern, W. M. 2011. "Darrell A. Posey (1947–2001)." *Tipití: Journal of the Society for the Anthropology of Lowland South America* 2 (1): 79–89.

Hershatter, G. 2011. *The Gender of Memory: Rural Women and China's Collective Past*. Berkeley: University of California Press.

Hinton, W. 1966. *Fanshen: A Documentary of the Revolution in a Chinese Village*. New York: Vintage.

Hinton, W. 1990. *The Great Reversal: The Privatization of China*. New York: Monthly Review Press.

Ho, F. W., and B. Mullen, eds. 2008. *Afro Asia: Revolutionary Political and Cultural Connections between African Americans and Asian Americans*. Durham, NC: Duke University Press.

Ho, Peter. 2001. "Greening without conflict? Environmentalism, NGOs and civil society in China." *Development and Change* 32 (5): 893–921.

Ho, Peter. 2003. "Mao's War against Nature? The Environmental Impact of the Grain-First Campaign in China." *China Journal* 50 (July): 37–59.

Hodgson, D. L. 2011. *Being Maasai, Becoming Indigenous: Postcolonial Politics in a Neoliberal World*. Bloomington: Indiana University Press.

Hodgson, Dorothy L. 2002. "Introduction: Comparative Perspectives on the Indigenous Rights Movement in Africa and the Americas." *American Anthropologist* 104 (4): 1037–49.

Hoy, D. C. 2004. *Critical Resistance: From Poststructuralism to Post-Critique*. Cambridge, MA: The MIT Press.

Hribal, J. 2011. *Fear of the Animal Planet: The Hidden History of Animal Resistance*. Oakland, CA: AK Press.

Hsu, Elisabeth, and Chris Low, eds. 2008. *Wind, Life, Health: Anthropological and Historical Perspectives*. Malden, MA: Wiley-Blackwell.

Hsueh, Chi-ju. 1985. "Reminiscences of Collecting the Type Specimens of Metasequoia Glyptostroboides H. H. Hu & Cheng." *Arnoldia* 45 (4): 10–18.

Hu Houxuan. 1944. "*Jiaguwen Sifang Fengming Kaozheng*" In his *Jiaguxue Shangshi Luncong* (Book of Essays on Oracle Bone Studies). 1–6. Chengdu: Chilu University.

Hyde, S. T. 2007. *Eating Spring Rice: The Cultural Politics of AIDS in Southwest China*. Berkeley: University of California Press.

Hyde, W. F., B. M. Belcher, and J. Xu. 2003. *China's Forests: Global Lessons from Market Reforms*. New York: Resources for the Future Press.

Igoe, J. 2004. *Conservation and Globalization: A Study of the National Parks and Indigenous Communities from East Africa to South Dakota*. Belmont, CA: Thomson/Wadsworth.

Igoe, J. 2006. "Becoming Indigenous Peoples: Difference, Inequality, and the Globalization of East African Identity Politics." *African Affairs* 105 (420): 399–420.

Inda, J. X., and R. I. Rosaldo, eds. 2002. *The Anthropology of Globalization: A Reader*. New York: Blackwell.

International Society of Ethnobiology. 2009. "Annex 2: Outline of the Global Coalition as Prescribed by the Kunming Action Plan," http://ethnobiology.net/about/ise-constitution/ (accessed October 12, 2012).

Ives, J. D., and B. Messerli. 1989. *The Himalayan Dilemma: Reconciling Development and Conservation.* New York: Routledge.

Jackson, J. E., and K. B. Warren. 2005. "Indigenous Movements in Latin America, 1992–2004: Controversies, Ironies, New Directions." *Annual Review of Anthropology* 34: 549–73.

Jacoby, K. 2003. *Crimes against Nature: Squatters, Poachers, Thieves, and the Hidden History of American Conservation.* Berkeley: University of California Press.

Jefferess, D. 2008. *Postcolonial Resistance: Culture, Liberation and Transformation.* Toronto: University of Toronto Press.

Jepson, P, and S. Canney. 2003. "The State of Asian Elephant Conservation in 2003." Unpublished Report for Elephant Family (London).

Jung, C. 2008. *The Moral Force of Indigenous Politics: Critical Liberalism and the Zapatistas.* Cambridge, UK: Cambridge University Press.

Karl, R. E. 2002. *Staging the World: Chinese Nationalism at the Turn of the Twentieth Century.* Durham, NC: Duke University Press.

Keck, W., and K. Sikkink. 1998. *Activists beyond Borders: Advocacy Networks in International Politics.* Ithaca, NY: Cornell University Press.

Kelley, R. D. G., and B. Esch. 1999. "Black Like Mao: Red China and Black Revolution." *Souls* 1 (4): 6–41.

Kimball, R. 2001. *The Long March: How the Cultural Revolution of the 1960s Changed America.* San Francisco: Encounter Books.

Kingsbury, Benedict. 1995. "'Indigenous Peoples' as an International Legal Concept." In *Indigenous Peoples of Asia*, ed. R.H. Barnes, Andrew Gray, and Benedict Kingsbury, 13–35. Ann Arbor: Association for Asian Studies.

Kingsbury, Benedict. 1998. "'Indigenous Peoples' in International Law: A Constructivist Approach to the Asian Controversy." *American Journal of International Law* 92 (3): 414–57.

Kinzley, J. C. 2012. "Crisis and the Development of China's Southwestern Periphery: The Transformation of Panzhihua, 1936–1969." *Modern China* 38 (5): 559–584.

Kipnis, A. 1994. "(Re)inventing Li: Koutou and Subjectification in Rural Shandong." In *Body, Subject, and Power in China.* 201–223, Chicago: University of Chicago Press.

Kipnis, Andrew B. 1997. *Producing Guanxi: Sentiment, Self, and Subculture in a North China Village.* Durham, NC: Duke University Press.

Kipnis, Andrew B. 2008. "Audit Cultures: Neoliberal Governmentality, Socialist Legacy, or Technologies of Governing?" *American Ethnologist* 35 (2): 275–89.

Kirsch, S. 2007. "Indigenous Movements and the Risks of Counterglobalization: Tracking the Campaign against Papua New Guinea's Ok Tedi Mine." *American Ethnologist* 34 (2): 303–21.

Klooster, Dan. 2000. "Community Forestry and Tree Theft in Mexico: Resistance or Complicity in Conservation?" *Development and Change* 31: 281–305.

Kojevnikov, A. 2008. "The Phenomenon of Soviet Science." *Osiris* 23 (1): 115–35.

Kuriyama, Shigehisa. 1994. "The Imagination of Winds and the Development of the Chinese Conception of the Body." In *Body, Subject, and Power in China*, ed. Angela Zito and Tani Barlow, 23–41. Chicago: University of Chicago Press.

Lang, Graeme. 2002. "Forests, Floods, and the Environmental State in China." *Organization and Environment* 15 (2): 109–30.

Larkin, B. D. 1971. *China and Africa, 1949–1970: The Foreign Policy of the People's Republic of China*. Berkeley: University of California Press.

Larkin, B. D. 1975. "Chinese Aid in Political Context: 1971–73." In *Chinese and Soviet Aid to Africa*, ed. Warren Weinstein, 1–28. New York: Praeger.

Latour, Bruno. 1987. *Science in Action: How to Follow Scientists and Engineers through Society*. Cambridge, MA: Harvard University Press.

Latour, B., and S. Woolgar. 1979. *Laboratory Life: The Social Construction of Scientific Facts*. Beverley Hills: Sage.

Latour, Bruno. 1993. *The Pasteurization of France*. Cambridge, MA: Harvard University Press.

Leach, E. R. 1964. "Anthropological Aspects of Language: Animal Categories and Verbal Abuse." In *New Directions in the Study of Language*, ed. E. H. Lenneberg, 23–63. Cambridge, MA: The MIT Press.

Leach, Melissa, and Robin Mearns, eds. 1996. *The Lie of the Land: Challenging Received Wisdom on the African Environment*. Oxford: James Currey.

Lee, Christopher J, ed. 2010. *Making a World after Empire: The Bandung Moment and Its Political Afterlives*. Athens: Ohio University Press.

Lehane, Robert. 1993. "ACIAR Ecological Economics Project 1993/105." Unpublished Report by the Australian Centre for International Agricultural Research (ACIAR).

Lévi-Strauss, C. 1963. *Totemism*. Boston: Beacon Press.

Lewis, M. L. 2004. *Inventing Global Ecology: Tracking the Biodiversity Ideal in India, 1947–1997*. Athens: Ohio University Press.

Li, A. 2011. "Cultural Heritage and China's Africa Policy." In *China and the European Union in Africa: Partners or Competitors?*, ed. Jing Men and Benjamin Barton, 41–60. Farnham, UK: Ashgate.

Li, Bo. 2001. "'The Lost Horizon': In Search of Community-Based Natural Resource Management in Nature Reserves of Northwest Yunnan, China." M.A. thesis, Cornell University.

Li, Gui, and Jiru Xue. 1995. *Gaoligong Mountain National Nature Reserve* (Gaoligongshan Guoji Ziran Baohuqu). Beijing: Chinese Forestry Publishing House (Zhongguo Linye Chubanshe).

Li, H., T. M. Aide, Y. Ma, W. Liu, and M. Cao. 2007. "Demand for Rubber Is Causing the Loss of High Diversity Rain Forest in SW China." *Biodiversity and Conservation* 16 (6): 1731–45.

Li, J. Y. 2010. "Life Science: Innovation and Prosperity—Commemorating the 60th Anniversary of the Chinese Academy of Sciences." *Science China* 53 (1): 2–12.

Li, P. J. 2007. "Enforcing Wildlife Protection in China." *China Information* 21 (1): 71–107.

Li, T. 2000. "Articulating Indigenous Identity in Indonesia: Resource Politics and the Tribal Slot." *Comparative Studies in Society and History* 42 (1): 149–79.

Li, T. 2007. *The Will to Improve: Governmentality, Development, and the Practice of Politics.* Durham, NC: Duke University Press.

Li, Wenhua, and Xianying Zhao. 1990. *China's Nature Reserves.* Beijing: Foreign Language Press.

Li, Xiwen, and D. Walker. 1986. "The Plant Geography of Yunnan Province, Southwest China." *Journal of Biogeography* 13 (5): 367–97.

Li, Zhinan. 2006. "The Loss of Shifting Cultivation: From Development Target to Agent." Paper Presented at the conference, Regionalization of Development: Redefining Local Culture, Space and Identity in The Mekong Region. April 22–24. Luang Prabang, Lao PDR.

Li, Zhixiang, Shilai Ma, Chenghui Hua, and Yingxiang Wang. 1982. "The Distribution and Habits of the Yunnan Golden Monkey, Rhinopithecus Bieti." *Journal of Human Evolution* 11 (7): 633–38.

Liang, L., L. Shen, W. Yang, X. Yang, and Y. Zhang. 2009. "Building on Traditional Shifting Cultivation for Rotational Agroforestry: Experiences from Yunnan, China." *Forest Ecology and Management* 257 (10): 1989–94.

Lieberman, S. T. 1991. "Visions and Lessons: 'China' in Feminist Theory-Making, 1966–1977." *Michigan Feminist Studies* 6: 91–107.

Liebes, T., and E. Katz. 1990. *The Export of Meaning.* Oxford: Oxford University Press.

Liou, Caroline, Marie Cambon, Alexander English, Thomas Huhti, Korina Miller, and Bradley Wong. 2000. *Lonely Planet: China.* Oakland, CA: Lonely Planet Publications.

Litzinger, R. 2004. "The Mobilization of Nature: Perspectives from North-west Yunnan." *The China Quarterly* 178: 488–504.

Litzinger, R. A. 2006. "Contested Sovereignties and the Critical Ecosystem Partnership Fund." *PoLAR: Political and Legal Anthropology Review* 29 (1): 66–87.

Litzinger, Ralph. 2000. *Other Chinas: The Yao and the Politics of National Belonging.* Durham, NC: Duke University Press.

Liu, L. H. 1995. *Translingual Practice: Literature, National Culture, and Translated Modernity—China, 1900–1937.* Stanford: Stanford University Press.

Liu, Dachang. 2001. "Tenure and Management of Non-State Forests in China since 1950: A Historical Review." *Environmental History* 6 (2): 239–63.

Liu, Hong. 1998. "Old Linkages, New Networks: The Globalization of Overseas Chinese Voluntary Associations and Its Implications." *China Quarterly* 155: 582–609.

Liu, Hongmao, Zaifu Xu, Youkai Xu, and Jinxiu Wang. 2002. "Practice of Conserving Plant Diversity through Traditional Beliefs: A Case Study in Xishuangbanna, Southwest China." *Biodiversity and Conservation* 11 (4): 705–13.

Liu, X. 2000. *In One's Own Shadow: An Ethnographic Account of Post-Reform Rural China.* Berkeley: University of California Press.

Liu, Zhiqiu. 2006. *Changes of Agrobiodiversity in Yakuo (Yunnan) during the Last Five Decades—With Focus on the Diversity of Rice.* Indigenous Knowledge

Program, Working Paper 5. Kunming, China: Center for Biodiversity and Indigenous Knowledge.

Long, C. L., and Y. Zhou. 2001. "Indigenous Community Forest Management of Jinuo People's Swidden Agroecosystems in Southwest China." *Biodiversity and Conservation* 10 (5): 753–67.

López-Pujol, Jordi, Fu-Min Zhang, Hai-Qin Sun, Tsun-Shen Ying, and Song Ge. 2011. "Mountains of Southern China as 'Plant Museums' and 'Plant Cradles': Evolutionary and Conservation Insights." *Mountain Research and Development* 31 (3): 261–69.

Lora-Wainwright, Anna. 2013. *Fighting for Breath: Living Morally and Dying of Cancer in a Chinese Village*. Honolulu: University of Hawaii Press.

Loss, Scott R., Tom Will, and Peter P. Marra. 2013. "The Impact of Free-ranging Domestic Cats on Wildlife of the United States." *Nature Communications* 4 (2013): 1396.

Lowe, C. 2006. *Wild Profusion: Biodiversity Conservation in an Indonesian Archipelago*. Princeton, NJ: Princeton University Press.

Lu, Junpei, and Qingbo Zeng. 1981. "A Preliminary Observation on the Ecological Consequence After 'Slash-and-Burn Cultivation' of the Tropical Semideciduous Monsoon Forest on the Jian Feng Mountain in Hainan Island." *Acta Phytoecologica Sinica* 4: 271–280.

Lu, S. C. 2008. "'Trade with the Devil': Rubber, Cold War Embargo, and U.S.-Indonesian Relations, 1951–1956." *Diplomacy and Statecraft* 19 (1): 42–68.

Luke, T. 1999. "Environmentality as Green Governmentality." In *Discourses of the Environment*, ed. Éric Darier, 121–51. Malden, MA: Blackwell Publishers.

Luo, Yiqun. 2008. "Local Ecological Knowledge of the Miao Ethnic Minority and the Restoration and Renewal of Forest Ecosystem." *Journal of Tongren University* 6: 12–17.

Ma Jincong. 2003. "Shuishan Wei Jie Mi de Chutan" (A preliminary study of Metasequoia). *Yunnan zhiwu yanjiu* (Yunnan Botanical Research) 25 (2): 155–72.

Ma, Shilai, Lianxian Han, and Daoying Lan. 1994. "Bird and Mammal Resources, and Nature Conservation in the Gaoligongshan Region, Yunnan Province, PRC." Unpublished Report for the Kunming Institute of Zoology.

MacKinnon, John. 1974. *In Search of the Red Ape*. Cambridge, UK: Cambridge University Press.

MacKinnon, John. 1991. The Story of Xishuangbanna. Film (52 mins). Hong Kong: World Wildlife Fund-China. In author's possession.

MacKinnon, John, and Karen Phillips. 2000. *A Field Guide to the Birds of China*. Oxford: Oxford University Press.

MacKinnon, John, Meng Sha, Catherine Cheung, Geoff Carey, Zhu Xiang, David Melvile, and Geoff Carey. 1996. *A Biodiversity Review of China*. Hong Kong: World Wide Fund for Nature International.

Maeda, D. J. 2006. "Black Panthers, Red Guards, and Chinamen: Constructing Asian American Identity through Performing Blackness, 1969–1972." *American Quarterly* 57 (4): 1079–103.

Mahmood, S. 2005. *Politics of Piety: The Islamic Revival and the Feminist Subject.* Princeton, NJ: Princeton University Press.
Marcuse, Gary. 2011. *Waking the Green Tiger: The Rise of the Green Movement in China.* Film (78 min). Vancouver: Video Project.
Marks, Robert. 2012. *China: Its Environment and History.* Lanham, MD: Rowman & Littlefield.
Massey, D. B. 2005. *For Space.* London: Sage.
Matsuzawa, Setsuko. 2011. "Horizontal Dynamics in Transnational Activism: The Case of Nu River Anti-Dam Activism in China." *Mobilization: An International Quarterly* 16 (3): 369–87.
Matsuzawa, Setsuko. 2012. "Citizen Environmental Activism in China: Legitimacy, Alliances, and Rights-based Discourses." *ASIANetwork Exchange: A Journal for Asian Studies in the Liberal Arts* 19 (2) : 81–91.
Matthiessen, P. 2001. *The Birds of Heaven: Travels with Cranes.* New York: Farrar, Straus & Giroux.
Mavhunga, C. C. 2011. "A Plundering Tiger with Its Deadly Cubs? The USSR and China as Weapons in the Engineering of a 'Zimbabwean Nation,' 1945–2009." In *Entangled Geographies: Empire and Technopolitics in the Global Cold War*, ed. G. Hecht, 231–66. Cambridge, MA: The MIT Press.
McFadyen, R. C. 2003. "Chromolaena in Southeast Asia and the Pacific." In *Agriculture: New Directions for a New Nation- East Timor (Timor-Leste)*, ed. H. da Costa, C. Piggin, C. J. da Cruz, and J. J. Fox, 130–34. Bruce, Australia: Australian Centre for International Agricultural Research.
McKee, K. 2009. "Post-Foucauldian Governmentality: What Does It Offer Critical Social Policy Analysis?" *Critical Social Policy* 29 (3): 465–86.
McKeown, A. 2007. "Periodizing Globalization." *History Workshop Journal* 63: 218–30.
McNeill, John. 1994. "Of Rats and Men: A Synoptic Environmental History of the Island Pacific." *Journal of World History* 5 (2): 299–349.
Meggers, Betty J. 1971. *Amazonia: Man and Nature in a Counterfeit Paradise.* Chicago: Aldine.
Menzies, N. K. 1994. *Forest and Land Management in Imperial China.* London: St. Martin's Press.
Merry, S. E. 2006. *Human Rights and Gender Violence: Translating International Law into Local Justice.* Chicago: University of Chicago Press.
Mertha, A. 2008. *China's Water Warriors: Citizen Action and Policy Change.* Ithaca, NY: Cornell University Press.
Miller, D. 1995. "Consumption and Commodities." *Annual Review of Anthropology* 141–61.
Mintz, Sidney W. 1985. *Sweetness and Power: The Place of Sugar in Modern History.* New York: Viking Penguin.
Mitchell, T. 1988. *Colonising Egypt.* Berkeley: University of California Press.
Mitchell, T. 1990. "Everyday Metaphors of Power." *Theory and Society* 19 (5): 545–77.

Mol, A. P., and F. H. Buttel, eds. 2002. *The Environmental State under Pressure*. Oxford: Elsevier Science.

Monson, Jamie. 2009. *Africa's Freedom Railway: How a Chinese Development Project Changed Lives and Livelihoods in Tanzania*. Bloomington: Indiana University Press.

Moore, D. S. 2000. "The Crucible of Cultural Politics: Reworking Development in Zimbabwe's Eastern Highlands." *American Ethnologist* 26: 654–89.

Moore, M. 2009. "China's Middle-Class Rise Up in Environmental Protest." *The Telegraph*, London, November 23.

Moore, Malcolm. 2011. "China Bans Animal Circuses." *The Telegraph*. London, January 18.

Morris, Ramona, and Desmond Morris. 1966. *Men and Pandas*. New York: McGraw-Hill.

Mueggler, Erik. 1998. "The Poetics of Grief and the Price of Hemp in Southwest China." *Journal of Asian Studies* 57: 979–1008.

Mueggler, Erik. 2001. *The Age of Wild Ghosts: Memory, Violence, and Place in Southwest China*. Berkeley: University of California Press.

Mueggler, Erik. 2002. "Dancing Fools: Politics of Culture and Place in a 'Traditional Nationality Festival.'" *Modern China* 28 (1): 3–38.

Mueggler, Erik. 2011. *The Paper Road: Archive and Experience in the Botanical Exploration of West China and Tibet*. Berkeley: University of California Press.

Muehlebach, A. 2001. "'Making Place' at the United Nations: Indigenous Cultural Politics at the U.N. Working Group on Indigenous Populations." *Cultural Anthropology* 16 (3): 415–48.

Mullaney, T. S. 2004. "Ethnic Classification Writ Large: The 1954 Yunnan Province Ethnic Classification Project and Its Foundations in Republican-Era Taxonomic Thought." *China Information* 18 (2): 207–41.

Mullin, M. 1999. "Mirrors and Windows: Sociocultural Studies of Human-Animal Relationships." *Annual Review of Anthropology* 201–24.

Mullin, M. 2002. "Animals and Anthropology." *Society and Animals* 10 (4): 387–94.

Murphy, Michelle. 2012. *Seizing the Means of Reproduction: Entanglements of Feminism, Health, and Technoscience*. Durham, NC: Duke University Press.

Muscolino, M. S. 2009. *Fishing Wars and Environmental Change in Late Imperial and Modern China*. Cambridge, MA: Harvard University Council on East Asian Studies.

Myers, N. 1988. "Threatened Biotas: 'Hot Spots' in Tropical Forests." *Environmentalist* 8 (3): 187–208.

Myers, Norman. 1984. *The Primary Source: Tropical Forests and Our Future*. New York: Norton.

Nair, P.K.R. 1996. "Agroforestry Directions and Literature trends." In *The Literature of Forestry and Agroforestry*, ed. Peter McDonald and James Lassoie, 74–95. Ithaca, NY: Cornell University Press.

The Nature Conservancy. 2003. *The Field Guide to the Nature Conservancy*. Arlington, VA: TNC.

Nedostup, R. 2009. *Superstitious Regimes*. Cambridge, MA: Harvard University Asia Center.

Neumann, Roderick P. 1998. *Imposing Wilderness: Struggles over Livelihood and Nature Preservation in Africa*. Berkeley: University of California Press.

Neushul, P., and Z. Wang. 2000. "Between the Devil and the Deep Sea: C. K. Tseng, Mariculture, and the Politics of Science in Modern China." *Isis* 91 (1): 59–88.

Newman, Steven Mark, and Bert Seibert. 1995. *Malaysia, Vietnam, China: Evaluation of Conservation Strategies in Asian Countries*. Unpublished report for the World Wide Fund for Nature (Geneva).

Niezen, R. 2000. "Recognizing Indigenism: Canadian Unity and the International Movement of Indigenous Peoples." *Comparative Studies in Society and History* 42 (1): 119–48.

Niezen, Ronald. 2003. *The Origins of Indigenism: Human Rights and the Politics of Identity*. Berkeley: University of California Press.

Oakes, T. 1999. "Eating the Food of the Ancestors: Place, Tradition, and Tourism in a Chinese Frontier River Town." *Cultural Geographies* 6 (2): 123–45.

Oates, J. F. 1999. *Myth and Reality in the Rain Forest*. Berkeley: University of California Press.

O'Brien, K. J., and L. Li. 2006. *Rightful Resistance in Rural China*. Cambridge, UK: Cambridge University Press.

O'Brien, William. 2002. "The Nature of Shifting Cultivation: Stories of Harmony, Degradation, and Redemption." *Human Ecology* 30 (4): 483–502.

O'Hearn, D. 2009. "Repression and Solidarity Cultures of Resistance: Irish Political Prisoners on Protest." *American Journal of Sociology* 115 (2): 491–526.

Oliver, R. 1978. "Distribution and Status of the Asian Elephant." *Oryx* 14: 379–424.

Omi, M., and H. Winant. 1986. *Racial Formation in the United States: From the 1960s to the 1980s*. New York: Routledge and Kegan Paul.

Ong, A., and S. J. Collier, eds. 2005. *Global Assemblages: Technology, Politics, and Ethics as Anthropological Problems*. Malden, MA: Blackwell.

Ong, Aihwa. 1987. *Spirits of Resistance and Capitalist Discipline: Factory Women in Malaysia*. Albany: State University of New York Press.

Ortner, Sherry. 1995. "Resistance and the Problem of Ethnographic Refusal." *Comparative Studies in Society and History* 37 (1): 173–93.

Paloczi-Horvath, G. 1962. *Mao Tse-tung: Emperor of the Blue Ants*. Westport, CT: Greenwood Press.

Peet, Richard, and Michael Watts, eds. 1996. *Liberation Ecologies: Environment, Development, Social Movements*. London: Routledge.

Pei, Shengji, and Percy Sajise, eds. 1995. *Regional Study on Biodiversity: Concepts, Frameworks, and Methods*. Kunming: Yunnan University Press.

Pei, Shengji. 1985. "Some Effects of the Dai People's Cultural Beliefs and Practices upon the Environment of Xishuangbanna, Yunnan, China." In *Cultural Values and Human Ecology in Southeast China*, ed. Carl Hutterer, A. Terry

Rambo, and George Lovelace, 321–339. Ann Arbor: University of Michigan Press.

Pei, Shengji. 1993. "Managing for Biological Diversity Conservation in Temple Yards and Holy Hills: The Traditional Practices of the Xishuangbanna Dai Community, Southwest China." In *Ethics, Religion and Biodiversity: Relations between Conservation and Cultural Values*, ed. L. D. Hamilton, 118–32. Birmingham, UK: White Horse.

Pei, Shengji, Jianchu Xu, Sanyang Chen, and Chunlin Long, eds. 1997. *Xishuangbanna Lunxie Nongye Shengtai Xitong Shengwu Duoyangxing Yanjiu Lunwen Baogao Ji* (collected Research Papers on Biodiversity in Swidden Agroecosystems in Xishuangbanna). Kunming, China: Yunnan Education Publishers.

Peluso, Nancy L. 1993. "Coercing Conservation: The Politics of State Resource Control." *Global Environmental Change* 3 (2): 199–218.

Peluso, Nancy L. 1994. *Rich Forests, Poor People*. Berkeley: University of California Press.

Perdue, Peter. C. 1987. *Exhausting the Earth: State and Peasant in Hunan, 1500–1850*. Cambridge, MA: Harvard University Asia Center.

Philip, Duke of Edinburgh. 1987. "Conservation in China." *Environmentalist* 7 (4): 245–52.

Pieterse, Jan Nederveen. 1995. "Globalization as Hybridization." In *Global Modernities*, ed. Mike Featherstone, Scott Lash, and Roland Robertson, 45–68. London: Sage Publications.

Pigg, Stacy Leigh. 1992. "Inventing Social Categories through Place: Social Representations and Development in Nepal." *Comparative Studies in Society and History* 34 (3): 491–513.

Pigg, Stacy Leigh. 1996. "The Credible and the Credulous: The Question of 'Villager's Beliefs' in Nepal." *Cultural Anthropology* 11 (2): 160–201.

Pigg, Stacy Leigh. 2001. "Languages of Sex and AIDS in Nepal: Notes on the Social Production of Commensurability." *Cultural Anthropology* 16 (4): 481–541.

Piot, Charles. 1999. *Remotely Global: Village Modernity in West Africa*. Chicago: University of Chicago Press.

Place, S., ed. 1993. *Tropical Rainforests: Latin American Nature and Society in Transition*. Wilmington, DE: Scholarly Resources.

Pomeranz, K. 2000. *The Great Divergence: China, Europe, and the Making of the Modern*. Princeton, NJ: Princeton University Press.

Povinelli, E. A. 2002. *The Cunning of Recognition: Indigenous Alterities and the Making of Australian Multiculturalism*. Durham, NC: Duke University Press.

Prashad, V. 2002. *Everybody Was Kung Fu Fighting: Afro-Asian Connections and the Myth of Cultural Purity*. Boston: Beacon Press.

Prestholdt, Jeremy. 2007. *Domesticating the World: African Consumerism and the Genealogies of Globalization*. Berkeley: University of California Press.

Pun, N. 2005. *Made in China: Women Factory Workers in a Global Workplace*. Durham, NC: Duke University Press.

Purcell, T. W. 1998. "Indigenous Knowledge and Applied Anthropology: Questions of Definition and Direction." *Human Organization* 57 (3): 258–72.

Rahmani, S. 2006. "Anti-imperialism and Its Discontents: An Interview with Mark Rudd, Founding Member of the Weather Underground." *Radical History Review* (95): 115–27.

Ramirez, R. K. 2007. *Native Hubs: Culture, Community, and Belonging in Silicon Valley and Beyond*. Durham, NC: Duke University Press.

Ramos, Alcida Rita. 1998. *Indigenism: Ethnic Politics in Brazil*. Madison: University of Wisconsin Press.

Ramsey, S. R. 1987. *The Languages of China*. Princeton, NJ: Princeton University Press.

Rappaport, Joanne. 2005. *Intercultural Utopias: Public Intellectuals, Cultural Experimentation, and Ethnic Pluralism in Colombia*. Durham, NC: Duke University Press.

Redford, Kent H. 1991. "The Ecologically Noble Savage." *Cultural Survival* 15 (1): 46–48.

Rees, Helen. 2000. *Echoes of History: Naxi Music in Modern China*. New York: Oxford University Press.

Reisner, M. 1991. *Game Wars: The Undercover Pursuit of Wildlife Poachers*. New York: Viking.

Rénique, Gerardo. 2009. "Law of the Jungle in Peru: Indigenous Amazonian Uprising against Neoliberalism." *Socialism and Democracy* 23 (3) :117–35.

Ritvo, Harriet. 1987. *The Animal Estate: The English and Other Creatures in the Victorian Age*. Cambridge, MA: Harvard University Press.

Robertson, Paul. 1995. "Glocalization: Time-Space and Homogeneity-Heterogeneity." In *Global Modernities*, ed. Mike Featherstone, Scott Lash, and Roland Robertson, 25–44. London: Sage.

Roe, Emery M. 1991. "Development Narratives, or Making the Best of Blueprint Development." *World Development* 19 (4): 287–300.

Roediger, D. R. 1999. *The Wages of Whiteness: Race and the Making of the American Working Class*. Brooklyn: Verso Books.

Rofel, Lisa. 1992. "Rethinking Modernity: Space and Factory Discipline in China." *Cultural Anthropology* 7 (1): 93–114.

Rose, N., P. O'Malley, and M. Valverde. 2006. "Governmentality." *Annual Review of Law and Social Science* 2: 83–104.

Ross, K. 2010. "China and Women's Liberation: Re-assessing the Relationship through Population Policies." *Hecate* 36 (1): 117–42.

Rowland, Beryl. 1974. *Animals with Human Faces: A Guide to Animal Symbolism*. London: Allen & Unwin.

Rudd, M. 2010. *Underground: My Life with SDS and the Weathermen*. New York: Harper Paperbacks.

Said, Edward. 1978. *Orientalism*. New York: Vintage.

Saint-Pierre, C. 1991. "Evolution of Agroforestry in the Xishuangbanna Region of Tropical China." *Agroforestry Systems* 12: 159–76.

Santiapillai, Charles, and Peter Jackson. 1990. *The Asian Elephant: An Action Plan for Its Conservation.* Gland, Switzerland: IUCN/SSC.

Santiapillai, Charles, Zhu Xiang, Dongyong Hua, and Shengqin Zhong. 1991. *Distribution of Asian Elephant in Xishuangbanna, People's Republic of China.* Unpublished Report for WWF-China (Hong Kong).

Sauer, H. D. 1999. "The Yangtze Flood 1998: A Corrected Outcome" (Das Yangzi-Hochwasser 1998: Eine Korrigierte Bilanz) (in German). *Wirtschaftswelt China* 99 (1): 11–12.

Schaller, George B. 1993. *The Last Panda.* Chicago: University of Chicago Press.

Schein, Louisa. 2000. *Minority Rules: The Miao and the Feminine in China's Cultural Politics.* Durham, NC: Duke University Press.

Schmalzer, Sigrid. 2002. "Breeding a Better China: Pigs, Practices, and Place in a Chinese County, 1929–1937." *Geographical Review* 92 (1): 1–23.

Schmalzer, Sigrid. 2008. *The People's Peking Man: Popular Science and Human Identity in Twentieth-Century China.* Chicago: University of Chicago Press.

Schmalzer, Sigrid. 2009. "Speaking about China, Learning from China: Amateur China Experts in 1970s America." *Journal of American–East Asian Relations* 16 (4): 313–52

Scott, J. C. 1990. *Domination and the Arts of Resistance: Hidden Transcripts.* New Haven, CT: Yale University Press.

Scott, J. C. 2009. *The Art of Not Being Governed: An Anarchist History of Upland Southeast Asia.* New Haven, CT: Yale University Press.

Scott, James C. 1985. *Weapons of the Weak.* New Haven, CT: Yale University Press.

Scott, James C. 1998. *Seeing Like a State.* New Haven, CT: Yale University Press.

Secretariat of the Chinese National Committee for MAB. 2004. *The Report of Field Review in Xishuangbanna Biosphere Reserve.* Jinghong. In author's possession.

Shanklin, E. 1985. "Sustenance and Symbol: Anthropological Studies of Domesticated Animals." *Annual Review of Anthropology* 14: 375–403.

Shao, Jing. 2006. "Fluid Labor and Blood Money: The Economy of HIV/AIDS in Rural Central China." *Cultural Anthropology* 21 (4): 535–69.

Shapiro, J. 2001. *Mao's War against Nature: Politics and the Environment in Revolutionary China.* Cambridge, UK: Cambridge University Press.

Sharma, A. 2006. "Crossbreeding Institutions, Breeding Struggle: Women's Empowerment, Neoliberal Governmentality, and State (Re)Formation in India." *Cultural Anthropology* 21 (1): 60–95.

Shiva, Vandana. 1988. *Staying Alive: Women, Ecology, and Development.* London: Zed Books.

Shou, Z.H., Y.T. Gao, and C.K. Lu. 1959. "Yunnan nanbu de xiang." (The elephants of Southern Yunnan Province) *Dongwuxue Zazhi (Journal of Zoology)* 5: 204–09.

Showers, K. B. 2005. *Imperial Gullies: Soil Erosion and Conservation in Lesotho.* Athens: Ohio University Press.

Simpson, J., and E. Weiner, eds. 1991. *The Compact Oxford English Dictionary.* Oxford: Oxford University Press.

Slater, Candace. 2003. *In Search of the Rain Forest*. Durham, NC: Duke University Press.

Smil, V. 1984. *The Bad Earth: Environmental Degradation in China*. Armonk, NY: M. E. Sharpe.

Smil, Vaclav. 1993. *China's Environmental Crisis: An Inquiry into the Limits of National Development*. Armonk, NY: M. E. Sharpe.

Sobel, David. 1989. "Beyond Ecophobia." *YES!* Winter: 19–23.

Songster, E. E. 2001. "Cultivating the Nation in Fujian's Forests: Forest Policies and Afforestation Efforts in China, 1911–1937." *Environmental History* 8: 452–73.

Songster, E. E. forthcoming. *Panda Nation: Nature, Science, and Nationalism in the People's Republic of China*.

Soroos, Marvin S. 1998. "The Assault on Tropical Rain Forests." *Mershon International Studies Review* 42 (2): 317–321.

Sparke, M. 2008. "Political Geography: Political Geographies of Globalization III—Resistance." *Progress in Human Geography* 32 (1): 1–18.

Spence, M. D. 2000. *Dispossessing the Wilderness: Indian Removal and the Making of the National Parks*. Oxford: Oxford University Press.

Spitulnik, Debra. 1993. "Anthropology and Mass Media." *Annual Review of Anthropology* 293–315.

Stahler-Sholk, R., H. E. Vanden, and G. D. Kuecker. 2007. "Globalizing Resistance: The New Politics of Social Movements in Latin America." *Latin American Perspectives* 34 (2): 5–16.

Steger, M. B. 2004. *Rethinking Globalism*. Lanham, MD: Rowman & Littlefield.

Stott, Philip. 1999. "Tropical Rain Forest: A Political Ecology of Hegemonic Mythmaking." Working Paper. London: Institute of Economic Affairs.

Sturgeon, Janet C. 2005. *Border Landscapes: The Politics of Akha Land Use in China and Thailand*. Seattle: University of Washington Press.

Sturgeon, Janet C. 2007. "Pathways of 'Indigenous Knowledge' in Yunnan, China." *Alternatives: Global, Local, Political* 32 (1): 129–53.

Sturgeon, Janet, and Nicholas Menzies. 2006. "Ideological Landscapes: Rubber in Xishuangbanna, Yunnan, 1950 to 2007." *Asian Geographer* 25 (1–2): 21–37.

Sun, Lena H. 1993. "Can Giant Pandas Survive the Effort to Save Them? Human Greed, Rivalry Dim Species' Future." *Washington Post*. December 27.

Sylvain, Renee. 2002. "'Land, Water, and Truth': San Identity and Global Indigenism." *American Anthropologist* 104 (4): 1074–85.

Tang, Chenggui, Zhanqiang Wang, and Zhihuai Zhang. 1998. *Xishuangbanna Daizu Zizhizhou Linye Zhi* (Xishuangbanna Dai Autonomous Prefecture Forestry Gazeteer). Kunming: Yunnan Minzu Chubanshe (Yunnan Nationalities Publishing House).

Tang, Ke. 1972. "China's Stand on the Question of the Human Environment." *Peking Review,* June 16: 5–8.

Tang, Xiyang. 1987. *Living Treasures: An Odyssey through China's Extraordinary Nature Reserves*. New York: Bantam Books.

Taussig, Michael. 1987. *Shamanism, Colonialism, and the Wild Man: A Study in Terror and Healing*. Chicago: University of Chicago Press.

Thaxton, R. 2008. *Catastrophe and Contention in Rural China: Mao's Great Leap Famine and the Origins of Righteous Resistance in Da Fo Village*. Cambridge, UK: Cambridge University Press.

Thouless, C. R., and J. Sakwa. 1995. "Shocking Elephants: Fences and Crop Raiders in Laikipia District, Kenya." *Biological Conservation* 72 (1): 99–107.

Tilt, Bryan. 2009. *The Struggle for Sustainability in Rural China: Environmental Values and Civil Society*. New York: Columbia University Press.

Tong, Enzheng. 1989. "Morgan's Model and the Study of Ancient Chinese Society." *Social Studies in China* 10 (2): 182–205.

Trouillot, M. R. 2003. *Global Transformations: Anthropology and the Modern World*. New York: Palgrave Macmillan.

Trouillot, Michel-Rolph. 1991. "Anthropology and the Savage Slot." In *Recapturing Anthropology*, ed. Richard G. Fox, 17–44. Santa Fe, NM: School of American Research Press.

Tsing, Anna L. 1999. "Becoming a Tribal Elder and Other Green Development Fantasies." In *Transforming the Indonesia Uplands: Marginality, Power and Productions*, ed. Tania M. Li, 159–202. Amsterdam: Harwood Academic.

Tsing, Anna L. 2003. "Agrarian Allegory and Global Futures." In *Nature in the Global South*, ed. P. Greenough and A. L. Tsing, 124–69. Durham, NC: Duke University Press.

Tsing, Anna L. 2005. *Friction: An Ethnography of Global Connection*. Princeton, NJ: Princeton University Press.

Tsing, Anna L. 2007. "Indigenous Voice." In *Indigenous Experience Today*, ed. Marisol de la Cadena and Orin Starn, 33–68. New York: Berg.

Turnbull, Colin M. 1972. *The Mountain People*. New York: Simon & Schuster.

Turner, M. D., and P. J. Taylor. 2003. "Critical Reflections on the Use of Remote Sensing and GIS Technologies in Human Ecological Research." *Human Ecology* 31 (2): 177–82.

Tuxill, J. D., G. P. Nabhan, E. Drexler, and M. Hathaway. 1998. *Plants and Protected Areas: A Guide to "In Situ" Management*. London: Stanley Thornes.

Uyghur Human Rights Project. 2009. "United Nations Declaration on the Rights of Indigenous People and the Uyghurs of East Turkestan (also known as the Xinjiang Uyghur Autonomous Region or XUAR, People's Republic of China)." Working Paper, http://docs.uyghuramerican.org/final_UNDRIP.pdf (accessed January 17, 2013).

Van Cott, D. L. 2003. "Indigenous Struggle." *Latin American Research Review* 38 (2): 220–33.

Walsh, Ellen Rose. 2005. "From Nu Guo to Nu'er Guo: Negotiating Desire in the Land of the Mosuo." *Modern China* 31 (4): 448–86.

Wan, Yanhai. 2001. "Becoming a Gay Activist in Contemporary China." *Journal of Homosexuality* 40 (3–4): 47–64.

Wang, Huihai. 1979. "The Soil Moisture Status of Tropical Rain Forest and Their Relation to the Growth and Development of Rainforest Plants in Southern Yunnan." *Acta Botanica Yunnanica* 2: 6–13.

Wang, Huihan, Weijun Ma, Chunzhang Deng, and Dehou Li. 1982. "The Exploitation of Tropical Rainforest in Relation to Soil and Water Conservation in Southern Yunnan" (Diannan Redaiyulin de Kaifa Liyong yu Shuitubaochi de Xianghu Guanxi). *Scientia Silvae Sinicae Linye Kexue* (Forestry Science) 3: 34–42.

Wang, Jianhua. 1998. "Traditional Culture and Biodiversity Management of Mountain Ethnic Groups in Xishuangbanna of Yunnan: A Case Study of Mengsong Hani Community." M.A. thesis, Kunming Institute of Botany.

Wang, Z. J., and S. S. Young. 2003. "Differences in Bird Diversity between Two Swidden Agricultural Sites in Mountainous Terrain, Xishuangbanna, Yunnan, China." *Biological Conservation* 110 (2): 231–43.

Warren, K. B. 1998. *Indigenous Movements and Their Critics: Pan-Maya Activism in Guatemala*. Princeton, NJ: Princeton University Press.

Watson, James L., ed. 1997. *Golden Arches East: McDonald's in East Asia*. Stanford: Stanford University Press.

Watts, M. 2003. "Development and Governmentality." *Singapore Journal of Tropical Geography* 24 (1): 6–34.

Watts, Michael. 2001. "Development Ethnographies." *Ethnography* 2 (2): 283–300.

Weinstein, Warren, ed. 1975. *Chinese and Soviet Aid to Africa*. New York: Praeger.

Welland, Sasha Su-Ling. 2006. Experimental Beijing: Contemporary Art Worlds in China's Capital. Ph.D. thesis., University of California, Santa Cruz.

Weller, Robert P. 2006. *Discovering Nature: Globalization and Environmental Culture in China and Taiwan*. Cambridge, UK: Cambridge University Press.

West, P. C., and S. R. Brechin, eds. 1991. *Resident Peoples and National Parks: Social Dilemmas and Strategies in International Conservation*. Tucson: University of Arizona Press.

Whatmore, Sarah. 1999. "Hybrid Geographies: Rethinking the 'Human' in Human Geography." In *Human Geography Today*, ed. Doreen Massey, John Allen, and Philip Sarre, 22–39. London: Polity Press.

Whatmore, Sarah, and Lorraine Thorne. 1998. "Wild(er)ness: Reconfiguring the Geographies of Wildlife." *Transactions of the Institute of British Geographers* 23 (4): 435–54.

White, S. D. 1997. "Fame and Sacrifice: The Gendered Construction of Naxi Identities." *Modern China* 23 (3): 298–327.

White, Tyrene. 1994. "The Origins of China's Birth Planning Policy." In *Engendering China: Women, Culture, and the State*, ed. Christina Gilmartin, Gail Hershatter, Lisa Rofel, and Tyrene White, 250–78. Cambridge, MA: Harvard University Press.

Wiens, H. J. 1954. *China's March towards the Tropics*. Camden, CT: Shoestring Press.

Williams, Dee Mack. 2002. *Beyond Great Walls: Environment, Identity, and Development on the Chinese Grasslands of Inner Mongolia.* Stanford: Stanford University Press.

Wilshusen, Peter R. 2010. "The Receiving end of Reform: Everyday Responses to Neoliberalisation in Southeastern Mexico." *Antipode* 42 (3): 767–99.

Wilshusen, P. R., S. R. Brechin, C. L. Fortwangler, and P. C. West. 2002. "Reinventing a Square Wheel: Critique of a Resurgent 'Protection Paradigm' in International Biodiversity Conservation." *Society & Natural Resources* 15 (1): 17–40.

Wolch, Jennifer R, and Jody Emel, eds. 1998. *Animal Geographies: Place, Politics, and Identity in the Nature-Culture Borderlands.* New York: Verso Books.

Wolch, Jennifer, Jody Emel, and Chris Wilbert. 2002. "Reanimating Cultural Geography." In *Handbook of Cultural Geography*, ed. Kay Anderson, M. Domosh, S. Pile, and N. Thrift, 184–206. London: Sage.

Wolf, Eric. 1982. *Europe and the People without History.* Berkeley: University of California Press.

Wolin, R. 2010. *The Wind from the East: French Intellectuals, the Cultural Revolution, and the Legacy of the 1960s.* Princeton, NJ: Princeton University Press.

Wong, J. 1975. "Chinese Demand for South-East Asian Rubber, 1949–72." *China Quarterly* (63): 490–514.

Wong, S. T., and S. S. Han. 1998. "Whither China's Market Economy? The Case of Lijin Zhen." *Geographical Review* 88 (1): 29–46.

Wood, S. L., and F. Huang. 1986. "New Genus of Scolytidae (Coleoptera) from Asia." *Western North American Naturalist* 46 (3): 465–67.

World Bank. 2008. Annex 9: Safeguard Policy Issues. China: Sustainable Development in Poor Rural Areas Project, http://www.google.com/url?sa=t&rct=j&q=&esrc=s&source=web&cd=1&cad=rja&ved=0CDQQFjAA&url=http%3A%2F%2Fwww-wds.worldbank.org%2Fservlet%2FWDSContentServer%2FIW3P%2FIB%2F2008%2F09%2F25%2F000334955_20080925045403%2FRendered%2FPDF%2FIPP3150Box334062B0P09957101PUBLIC1.pdf&ei=Xy4cUbTfI6OjigKuhIGYAQ&usg=AFQjCNEocPMq9RwbbNaXCJFoum4T7NxskA&sig2=4NvLmTetQtafkk2Y6GChCQ (accessed February 3, 2013).

Wu, C. Y. 1965. "The Tropical Floristic Affinity of the Flora of China." *Chinese Science Bulletin* 1: 25–33.

Wu, Fengshi. 2009. "Environmental Politics in China: An Issue Area in Review." *Journal of Chinese Political Science* 14 (4): 383–406.

Wu, Fengshi. 2013. "Environmental Activism in Provincial China: Comparative Evidence from Guangdong and Guangxi." *Journal of Environmental Policy and Planning* 15 (1): 89–108.

WWF. 1993. "Xishuangbanna Agroforestry Development Project: A Strategy and Procedure for Implementation." Unpublished Report for WWF-China (Beijing).

WWF. 1996. "Xishuangbanna Agroforestry Development Project: A Cooperative Venture of the People's Republic of China and the World Wide Fund for Nature (WWF)." Unpublished Report for WWF-China (Beijing).

Wylie, Dan. 2008. *Elephant*. Chicago: Reaktion Books.
Xanthaki, Alexandra. 2007. *Indigenous Rights and United Nations Standards*. Cambridge, UK: Cambridge University Press.
Xie, Lei. 2012. *Environmental Activism in China*. London: Routledge.
Xinhua. 2011. Local Government in Southwest China Purchases Insurance Policy to Compensate Residents Attacked by Wild Animals, http://news.xinhuanet.com/english2010/china/2011-01/04/c_13676553.htm (accessed February 10, 2013).
Xu, Jianchu. 1990. "Research on a Traditional Agroecological System." Unpublished report for the Institute of Botany, Academia Sinica (Kunming).
Xu, Jianchu. 2004. "Rattan and Tea-Based Intensification of Shifting Cultivation by Hani Farmers in Southwestern China." In *Voices from the Forest: Integrating Indigenous Knowledge into Sustainable Farming*, ed. M. Cairns. 667–673. New York: Resources for the Future.
Xu, Jianchu, J. Fox, N. Lu, N. Podger, S. Leisz, and X. Ai. 1999. "Effects of Swidden Cultivation, State Policies, and Customary Institutions of Land Cover in a Hani Village, Yunnan, China." *Mountain Research and Development* 19 (2): 123–32.
Xu, Jianchu, Yanhui Li, Shengji Pei, Sanyang Chen, and Kanlin Wang. 1995. "Swidden-Fallow Succession in the Mengsong Area of Xishuangbanna, Yunnan Province, China." In *Regional Study on Biodiversity: Concepts, Frameworks, and Methods*, ed. Shengji Pei and Percy Sajise, 183–96. Kunming: Yunnan University Press.
Xu, Jianchu, and Stephen Mikesell. 2003. "Indigenous Knowledge for Sustainable Livelihoods and Resource Governance in the MMSEA Region." In *Landscapes of Diversity: Indigenous Knowledge, Sustainable Livelihoods, and Resource Governance in Montane Mainland Southeast Asia*, ed. Jianchu Xu and Stephen Mikesell, 3–22. Kunming: Yunnan Science and Technology Press.
Xu, Jianchu, Shengji Pei, and Sanyang Chen. 1995. "From Subsistence to Market-Oriented System and the Impacts on Agroecosystem Biodiversity." In *Regional Study on Biodiversity: Concepts, Frameworks, and Methods*, ed. S. Pei and P. Sajise, 73–87. Kunming: Yunnan University Press.
Xu, Jianchu, Shengji Pei, and Sanyang Chen. 1997. "Indigenous Swidden Agroecosystems in Mengsong Hani Community." In *Biodiversity Research in Swidden Agroecosystems in Xishuangbanna*, ed. S. Pei and J. Xu, 26–33. Kunming: Yunnan Education Publishers.
Yan, Hairong. 2003. "Neoliberal Governmentality and Neohumanism: Organizing Suzhi/Value Flow through Labor Recruitment Networks." *Cultural Anthropology* 18 (4): 493–523.
Yan, Yunxiang. 1996. *The Flow of Gifts: Reciprocity and Social Networks in a Chinese Village*. Stanford: Stanford University Press.
Yang, Bin. 2009. "'We Want to Go Home!' The Great Petition of the Zhiqing, Xishuangbanna, Yunnan, 1978–1979." *The China Quarterly* 198 (1): 401–21.
Yang, D. L. 1998. *Calamity and Reform in China: State, Rural Society, and Institutional Change since the Great Leap Famine*. Stanford: Stanford University Press.

Yang, Guobin. 2005. "Environmental NGOs and Institutional Dynamics in China." *The China Quarterly* 181: 46–66.

Yang, Guobin, and Craig Calhoun. 2007. "Media, Civil Society, and the Rise of a Green Public Sphere in China." *China Information* 21 (2): 211–36.

Yang, Lan, Ruilang Pan, and Shuzhen Wang. 1985. "Investigation of the Birds of Cultivated Land of Tea Trees and Rubber Trees in Xishuangbanna, Yunnan Province." *Zoological Research* 4: 353–60.

Yang, Mayfair. 1994. *Gifts, Favors, and Banquets: The Art of Social Relations in China*. Ithaca, NY: Cornell University Press.

Yang, Mayfair Mei-hui. 2011. "Postcoloniality and Religiosity in Modern China: The Disenchantments of Sovereignty." *Theory, Culture & Society* 28 (2): 3–45.

Yang Rungao. 1998. "Xishuangbanna Senlin Jianshao de Chengyin Ii Dui Ziran Dili Huanjing de Yingxiang de Fenxi Yanjiu" (Causes of Reduction of Forest in Xishuangbanna and Analytical Analysis of Its Impact on the Natural Geographic Environment). PhD thesis, Yunnan Normal University. Kunming, China.

Yang, Ting-shuo. 2007. "Ecological Knowledge of the Miao People in the Control of Rocky Desertification." *Journal of Guangxi University for Nationalities (Philosophy and Social Sciences Edition)* 3: 24–33.

Yang, Yuming, Kun Tian, Jiming Hao, Shengji Pei, and Yongxing Yang. 2004. "Biodiversity and Biodiversity Conservation in Yunnan, China." *Biodiversity and Conservation* 13 (4): 813–26.

Yao, S. 1989. "Chinese Intellectuals and Science: A History of the Chinese Academy of Sciences (CAS)." *Science in Context* 3 (2): 447–73.

Ye, C., L. Fei, F. Xie, and J. Jiang. 2007. "A New Ranidae Species from China: Limnonectes Bannaensis (Ranidae: Anura)." *Zoological Research* 28 (5): 545–50.

Yeh, Emily T. 2005. "Green Governmentality and Pastoralism in Western China: Converting Pastures to Grasslands." *Nomadic Peoples* 9 (1/2): 9–29.

Yeh, Emily T. 2007. "Tibetan Indigeneity: Translations, Resemblances, and Uptake." In *Indigenous Experience Today*, ed. Marisol De La Cadena and Orin Starn, 69–98. London: Berg.

Yeh, Emily T. 2009. Greening Western China: A Critical View. *Geoforum* 40 (5): 884–94.

Yin, Shaoting. 1991. *Yige Chongman Zhengyi de Wenhua Shengtai Tixi: Yunnan Daogeng Huozhong Yanjiu* (A Highly Controversial Cultural-Ecological System: Studies in Swidden Agriculture in Yunnan). Kunming: Yunnan Renmin Chubanshe (Yunnan People's Publishing House).

Yin Shaoting. 1992. "Jinuozu de Daogeng Huozhong: Jian Yu Yunnan Qita Daogeng Huozhong Minzu de Bijiao" (The Swidden Agriculture of the Jinuo Nationality: With a Comparison of other Swidden Agriculturalist Peoples in Yunnan). *Kokuritsu Minzokugaku Hakubutsukan Kenyu Hokoku* (Research Reports from National Museum of Ethnology, Osaka). 17 (2): 268–274.

Yin, Shaoting. 1996. "Research on Yunnan Swidden Agriculture." Unpublished Report. Rikkyo University, Japan.

Yin, Shaoting. 2001. *Ren Yu Shenlin* (People and Forests: Yunnan Swidden Agriculture in Human-Ecological Perspective). Kunming: Yunnan Education Publishing House.

Young, C. A. 2006. *Soul Power: Culture, Radicalism, and the Making of a U.S. Third World Left*. Durham, NC: Duke University Press.

Young, Nick. 2005. "What about Guizhou? Reflections on Cooperation Prospects by James Harkness." *China Development Brief,* http://www.chinadevelopmentbrief.com/node/99 (accessed March 21, 2012).

Yu, Xiaogang. 1993. "Protected Areas, Traditional Natural Resource Management Systems and Indigenous Women: Case Study in Xishuangbanna, PR China." M.A. thesis, Asian Institute of Technology. Bangkok, Thailand.

Yu, Xiaogang. 1994. "Protected Areas, Traditional Natural Resource Management Systems and Indigenous Women: Case Study in Xishuangbanna, P.R. China." Paper presented at the MacArthur Grantees Meetings ICIMOD Seminar on Indigenous Knowledge Systems and Biodiversity Management, Kathmandu, Nepal, April 13–15.

Yu, Xiaogang. 2001. "Shui Zhiling Lai Zi Lijiang Lashi Hai Liuyu De Baogao" (Lashi Hai: The Soul of Water). *Huaxiaren Wenhua Dili* (Chinese Cultural Geography) 4 (August): 34–65.

Zhan, M. 2009. *Other-Worldly: Making Chinese Medicine through Transnational Frames*. Durham, NC: Duke University Press.

Zhang, Keying, and Yiping Zhang. 1984. *The Effect of Deforestation over Xishuangbanna Area on Local Climate*. Beijing: China Meteorological Press.

Zhang, X. 1996. "The Vietnam War, 1964–1969: A Chinese Perspective." *Journal of Military History* 60: 731–62.

Zhong, Xinshe. 2001. "Yunnan Province, Xishuangbanna, Destroyed More Than 60,000 Seized Firearms" (Yunnansheng Xishuangbanna Zhou Jiao Xiaohui Liu Wan Yuzhi Qiangzhi), http://news.sohu.com/23/67/news144396723.shtml (accessed February 1, 2012).

Zhou, M., and H. Sun, eds. 2004. *Language Policy in the People's Republic of China: Theory and Practice Since 1949*. Boston: Kluwer Academic.

INDEX

Academy of Social Sciences, Chinese, 140
accommodation model (of globalization), 23–25, 91, 115
actant, defined, 155
activism. *See* environmental activism
Activists beyond Borders, 22–23
Africa: colonial, anticolonial movements in, 29–30, 156; indigeneity and, 117, 118, 119, 128; nature conservation in, 13–14, 67, 83, 161, 162, 167, 183
agency, animal/nonhuman, 154–58
agency, cumulative, 156–57, 182–83
agency, elephants. *See under* elephants
agriculture. *See* slash-and-burn agriculture; swidden agriculture
agroforestry, 34–35, 66, 94–95, 96–98, 101–103, 105–106, 160–61, 177
Agroforestry Handbook, WWF, 97, 98*fig11*
Amazon, 128, 131
Amnesty International, 170
animals. *See* elephants; pandas; wildlife
Anti-Rightist Campaign/Movement, 48, 159
Apel, Ulrich, 129–30
Appadurai, Arjun, 22
art of engagement: *versus* arts of resistance, 73, 78–80, 88, 107; Chinese officials and, 88–90; compared to *guanxi xue*, 80, 205n3; environmental winds and, 38, 39, 94, 113–14, 115; expatriate conservationists and, 90–96, 100–101, 104–7; villagers and, 80–85, 88–91, 100–112
arts of resistance (Scott), 78–79, 205n3

Asian Development Bank (ADB), 10, 46, 119
Asian Elephant Specialist Group, 165
Asian Institute of Technology, Thailand, 143
Australia, 134, 137

Bamford, Dave, 162–63, 179
Banna. *See* Xishuangbanna (Banna)
Bareis, Karl, 1
Bentley, Jack: art of engagement, 90–96, 100–101, 104–7; promotion of agroforestry, 96–100, 101–3, 209n27; relationship to officials, Nature Reserve Bureau, 95–96, 102–3, 105, 106, 113–14, 209n25; transnational work, 63, 65–72; views on ethnic minorities, 97–99; villagers' engagement with, 108–12, 113–14
biodiversity: swidden agriculture and, 136–37, 141, 148; Yunnan as hotspot of, 10, 56
Black Maoism, 29
Black Panthers, 29, 30*fig4*, 186, 195n32
Black Power movement, globalized, 29, 33, 185
Buddhism/Buddhist, 91, 131, 134

Callon, Michel, 155–56
Canada, 7, 134, 137
capitalism: as logic of globalization, 6, 22, 24–25, 37, 86; socialist critique of, 8, 50, 122

251

Center for Biodiversity and Indigenous Knowledge (CBIK), 135, 140
Chadwick, Douglas, 164
Chen Zongyi, 52
China, People's Republic of (PRC), 9*map1*, 34*map2*; ethnological survey, 122; influence on global 60s, 27–33, 31*figs*, 32*fig6*, 36, 185–87; U.S.-led embargo against, 49, 51, 192n10, 199n5, 201n17. *See also* ethnic formations; ethnic minorities; political movements/campaigns (in China)
Chinese experts and scientists: as agents of environmental winds, 4–6, 11–12, 14–15, 21–22; defined, 189n2; indigeneity, human rights, and, 14, 118, 120, 125–26; persecution, rehabilitation of, 3, 43, 48–49, 50, 200n16; survey of wild elephants, 158–59; transnational work, 47–57, 70–72. *See also* Pei Shengji; Xu Jianchu; Yang Bilun; Yang Yuanchang; Yu Xiaogang
Chinese metaphor of winds. *See* winds *(feng)*, Chinese metaphor of
Chinese officials: as agents of environmental winds, 5–6, 11–12, 21–22, 193n14; art of engagement and, 88–90; negotiations with WWF, 44, 53, 58; population pressure/control, 60–61. *See also* Nature Reserve Bureau (Xishuangbanna)
CITES, 164, 165
civil rights movements, globalized, 27–33, 30*fig4*, 185
Cold War, 14, 49, 53–54, 100, 198n6
communes, 13, 16*fig1*, 17, 51, 91, 92, 171–72
communism/communist, 122–23, 142, 143
Communist Manifesto, 122
Communist Party, Chinese, 122, 134, 143
conservation, nature
—China's changing approaches to science and nature: environmentalism today, 8, 159; Mao era, 47–50, 69, 137, 147; Reform era, 50–53; wildlife and wetlands, 1950s–1980s, 15–19
—elephants and global networks of, 164–66, 183
—globalized trends in: community-based approaches, 19–22, 20*fig2;* necessity of linking into, 70–72; social forestry, 5, 19, 20*fig2*, 55
—as neo/colonialism, 13–14, 82–83, 206n11
—resistance to, 82–84
—*See also* agroforestry; biodiversity; environmentalism; slash-and-burn agriculture; swidden agriculture; tropical rain forests
Conservation International, 56, 128, 205n4
conservationists, expatriate
—as agents of environmental winds, 5–6, 11–12, 21–22
—defined, 190n4
—transnational work, 63–72
—views on ethnic minorities: as ecologically destructive, 61–63, 67, 99, 126, 135, 138; as possessors of knowledge, 34–35, 98–99, 114, 136, 166–67
—*See also* Bentley, Jack; MacKinnon, John; WWF (World Wildlife Fund)
Convention on International Trade in Endangered Species (CITES), 164, 165
Cuba, 150
Cultural Revolution, the, 10–11; described as a wind, 12–13, 191n4; impact on environment, 92, 159; impact on scientists, 50, 55, 64; impact on the global 60s, 27–33, 31*figs*, 32*fig6*, 36, 185–86
Cultural Values and Human Ecology in Southeast Asia, 128
cumulative agency, 156–57, 182–83

Dai, 62, 91, 104–5, 122, 129–31, 138, 139*fig12*, 212n12
Dave, Naisargi, 144
deforestation. *See under* slash-and-burn agriculture
Destroy the Four Pests Campaign, 15–16, 192n9
dingzi, 145, 146, 147
Domesticating the World: African Consumerism and the Genealogies of Globalization, 156
Domination and the Arts of Resistance, 78–79

Dragon Hills. *See* Holy Hills (or Dragon Hills)
DuBois, W. E. B., 29

Earth Day (1970), 10, 31
Earth First!, 10
East Wind *(Dong Feng)*, the, 26, 29, 32, 187, 194n27
ecological hotspots (Myers), 56, 57
ecotourism, 18, 160, 162–64. *See also* panda project, Sichuan (WWF); Wild Elephant Valley (WWF)
Egypt, 66
elephants: agency, 152–58, 161–62, 168–69, 175–79, 182–83; aggression of, 40, 152–53, 158, 175–77, 214n1; compensation for damages by, 89, 96, 105, 178–79, 180–81, 216n16–18, 217n24; death penalty for poaching, 170–71; defenses against, 160–62; environmental winds and, 152; importation of, 164; ivory trade, 48, 152, 183; murder rate (from), 152–53, 180, 217n23, 217n26; protecting, 173–75, 182; size of population (Banna), 153, 183, 216n13–14; survey of wild (first), 158–59; transnational networks and, 6, 48, 101, 115, 153–54, 164–65, 183–84, 215n4; villagers' admiration for, 85, 162, 178–79. *See also* guns, confiscation of; Wild Elephant Valley (WWF)
Emei Shan (sacred land), 134
endangered species. *See under* wildlife
engagement. *See* art of engagement
Engels, Friedrich, 122
environmental activism, 214n23–25; Yu Xiaogang's, 140–48. *See also* Pei Shengji; Xu Jianchu
environmentalism: China's advocacy for, 8, 159; critique of, 206n6; as global formation, 13–14, 37–38, 187, 190n7; imbrication of indigeneity and, 117–21, 125–26, 131, 135–40, 148–49. *See also* conservation, nature
environmental protests/petitions, 8, 51, 146, 214n26
environmental winds *(huanjing feng)*: agents of, 5–6, 11–12, 21–22; concepts that elucidate, 38; defined in contrast to global flows, 6–7, 11–12, 21–22, 36–37; environmentalism and, 3–4, 18, 36–38, 159, 191n3; as extension of idiom of winds, 13–15, 37; globalization and, 25–27, 36–37; Xiao Long and, 79–80, 112–15. *See also* globalized formations; winds *(feng)*, Chinese metaphor of
ethnic formations, 120–25, 211n6
ethnic majority. *See* Han Chinese
ethnic minorities *(shaoshu minzu)*
—expatriate conservationists' views on: as ecologically destructive, 61–63, 67, 99, 126, 135, 138; as possessors of knowledge, 14, 34–35, 98–99, 114, 136, 166–67
—lowland *versus* upland, 61–63, 130–31, 138, 171, 207n15
—named groups: Dai, 62, 91, 104–5, 122, 129–31, 138, 139*fig12*, 212n12; Jingpo, 64; Jinuo, 74, 174, 205n1, 206n8, 207n15; Manchu, 122; Mongol, 122; Naxi, 133–34; Tibetan, 118, 122, 125–26, 210n33, 211n10; Uyghurs, 118, 210n33, 211n7; Yi, 122
—splittism *(fen lie)*, 126, 203n33
ethnic tourism, 123
ethnological survey, China's, 122
European Union (EU), 33, 58, 66, 70, 71, 166, 167
expatriate conservationists. *See* conservationists, expatriate
experts. *See* Chinese experts and scientists

Faier, Lieba, 25
Fanon, Frantz, 29
Fanshen, 27
farmers. *See* villagers *(nongmin)*
farmers' rights, 35, 52
Fate of the Elephant, The, 164
feminism, globalized, 27–33, 36–38, 28*fig3*, 32*fig6*, 185, 198n47
feng. *See* winds *(feng)*
fengjian mixin, defined, 133. *See* superstition *versus* knowledge
Ford Foundation, 140, 141
Foreman, Dave, 10
forest guards. *See under* Nature Reserve Bureau (Xishuangbanna)

forestry, social, 5, 19, 20*fig2*, 55
Forestry Bureau (Xishuangbanna), 90, 100, 161, 163, 166
Forestry Department (Yunnan), 140
forests. *See* tropical rain forests
Foucault, Michel, 24, 194n21, 196n40
France, 30, 47
Fu Ganxing (Xiao Long villager), 108–19

Gaoligong Mountain National Nature Reserve, The, 2–3
gay rights, India, 144
Global Coalition for Biological and Cultural Diversity, 131
globalization: accommodation model, 23–25, 91, 115; critique of dominant models of, 6–7, 8–9, 24–25, 156–57, 186–87; indigenous resistance and, 23–24, 73, 79, 82, 86, 205n2–3, 206n11; localization model, 23–25, 91, 115; "making the global," 14–15, 37–40; resistance model, 23–25, 78–79, 85–88, 91; theories of, 22–26. *See also* environmental winds *(huanjing feng)*; globalized formations
globalized formations, 9–10, 22–27, 36–38, 191n2; civil rights movements, 27–33, 30*fig4*, 185; environmentalism, 13–14, 37–38, 187, 190n7; feminism, 27–33, 36–38, 28*fig3*, 32*fig6*, 185–86, 198n47; gay rights, 144; indigeneity, 115, 117–21, 187; socialism, 26, 30, 122, 150, 187
Goldman Prize, 146
governmentality (Foucault), 24, 194n22
"grain first" policy, 4
Great Leap Forward, the, 12, 13, 18–19, 50, 133, 191n5, 194n27
Greenhalgh, Susan, 60
GTZ (Gesellschaft für Technische Zusammenarbeit), 81, 129, 130
Guangdong (Xiao Long villager), 111–12
guanxi networks, 68, 84, 103, 111
guanxi xue, compared to art of engagement, 80, 205n3
Guha, Ranajit, 85, 86
guns, confiscation of, 8, 18–19, 40, 83, 152, 169, 173–75, 217n21

Han Chinese, 121–24, 125, 127, 129, 131, 133–34, 143, 205n1
Hanisch, Carol, 58
Hannerz, Ulf, 22
He Wei (Michael Hathaway), 179, 206n8
HIV/AIDS, 78, 145
Holy Hills (or Dragon Hills), 129, 136, 141. *See also* sacred lands
Hong Kong, 2, 44, 55, 137, 141
huanjing feng. See environmental winds *(huanjing feng)*
Huaxiaren Wenhua Dili, 143
Hu Jintao, 147
human-nonhuman relations, 154–57, 215n9

ICIMOD, 132, 141
India: environmentalism and, 24, 35, 67, 103, 129; resistance, indigeneity, 85, 87, 121; transnational connections with, 5, 48, 128, 132, 144, 164, 165, 166, 181
indigeneity: as constructed category, 125, 149–51; environmentalism and, 125–26, 136–40, 149–50; as globalized formation, 115, 117–21, 187; international framework for, 129; redefining, 39–40; rights and, 117–21, 125–26, 131; terminology of, 116, 120, 125, 132–34, 140, 143, 145, 148, 210n1, 211n3. *See also* Pei Shengji; Xu Jianchu; Yu Xiaogang
indigenous peoples and knowledge: advocating concepts of, 14, 140–45; resistance to globalization, 23–24, 73, 79, 82, 86, 205n2–3, 206n11; romantic notions of, 121, 211n4; superstition and, 132–34, 137, 148, 212n14–15. *See also* ethnic minorities *(shaoshu minzu)*; villagers *(nongmin)*
indigenous space, an, 118, 148–51, 210n2, 211n6–10. *See also* Pei Shengji; Xu Jianchu; Yu Xiaogang
Indonesia, 128, 165
International Centre for Integrated Mountain Development (ICIMOD), 132, 141
International Ethnobiology Meeting (2nd), 131–32
International Monetary Fund (IMF), 46

International Union for the Conservation of Nature, 165
Italy, 5
ivory trade, 48, 152, 183

Jackson, Peter, 165
Japan, 18, 47, 48, 55, 156
Jefferess, David, 86, 87
Jiang Zemin, 142, 143
Jingpo, 64
Jinuo, 74, 174, 205n1, 206n8, 207n15

Karl, Rebecca, 25
Keck, Margaret, 22
Korean War, 49, 54, 100
Kunming, China, as environmental hub, 41, 48, 140
Kunming Action Plan, 131–32
Kunming Institute of Botany, 132

Lao De (Xiao Long villager), 84, 114
Laos, 33, 48, 66, 128, 150, 164, 165, 182
Latin America, 117, 127, 206n11
Latour, Bruno, 155–56
Let a Hundred Flowers Bloom, 12, 200n15
Li, Lianjiang, 143
Li, Tania, 124
Li Baiwen (Xiao Long villager), 169–70, 180
Li Bo (Xiao Long villager), 109–10
Li Ming, 185
Little Red Book (Mao's), 12, 29
Living Treasures: An Odyssey through China's Extraordinary Nature Reserves, 61
Li Wen (Xiao Long villager), 105, 106
localization model (of globalization), 23–25, 91, 115
Lowe, Celia, 25
Lu Wen (Xiao Long villager), 108

Maasai, 130, 211n9
MAB program, 177, 180, 181
MacArthur Foundation, 141, 173
MacKinnon, John, 59–63, 65, 67, 69, 71, 99, 138, 202n31
Mahmood, Saba, 87
majority group. *See* Han Chinese
"making the global," 14–15, 37–40
Malcolm X, 29
Manchu, 122
Mao Zedong, 11, 26, 48, 49, 53, 142, 171; influence on global 60s, 27, 29, 30*fig4*, 33, 186–87
Marx, Karl, 122
Marxism, Marxist, 61, 133, 143, 150, 186
Ministry of Forestry, 70, 100, 182
minority groups. *See* ethnic minorities
minzu, defined, 122
mixin, defined, 132, 212n13–14. *See* superstition *versus* knowledge
Mongol, 122
Morgan, Henry Lewis, 122
Muir, John, 10
Myanmar, 33, 64, 128
Myers, Norman, 56, 57, 70, 203n34

nature conservation. *See* conservation, nature
Nature Reserve (Xishuangbanna), 179; creation of, 158–60
Nature Reserve Bureau (Xishuangbanna): art of engagement, 88–91, 113; attitudes toward indigenous peoples, 98, 141–42; forest guards, 4, 81, 82, 83, 84, 159; population relocation, 97, 141–42, 167; relations with WWF, 95–96, 102–3, 105, 106, 113–14; rising status/power of, 81, 83, 95–96, 178–80; transnational work, 68, 70, 179–80; villagers and, 81–82, 92, 102–103, 105–106, 169, 178, 182
nature tourism. *See* ecotourism
Naxi, 133–34
neoliberal/neoliberalism, 22, 23, 24, 186
Nepal, 5, 132, 194n23, 203n39
Newton, Huey, 29
New Zealand, 162
nomadic agriculture, 181

O'Brien, Kevin, 143
officials. *See* Chinese officials
"old stinking nines," 48, 200n16
Old Zhang (Xiao Long villager), 116, 119–20

Opening of China. *See* Reform and Opening Up *(gaige kaifang)*
Oxfam-Hong Kong, 141

panda project, Sichuan (WWF), 44–45, 59, 66, 71, 108, 182, 198n4
pandas, 15, 40, 44, 154, 165, 192n10, 216n16
peasants *(nongmin)*. *See* villagers *(nongmin)*
Pei Shengji, 127–32, 134, 137, 141, 147, 148, 149
People's Food Co-op (Ann Arbor, MI), 185–86
Peter Jackson, 165
Philip, HRH Prince, 44–45, 45*fig7*, 46, 47, 95
Philippines, 35, 63, 94, 128, 135, 147
Piot, Charles, 86
plantations: environmental impact of, 15, 51–52, 135–36, 166; rubber, 49–50, 69; tea, 78, 91–92
political movements/campaigns (in China)
—Anti-Rightist Campaign/Movement, 48, 159
—Cultural Revolution, the, 10–11; described as a wind, 12–13, 191n4; impact on environment, 92, 159; impact on global 60s, 27–33, 31*figs*, 32*fig6*, 36, 185–86; impact on scientists, 50, 55, 64; science *versus* superstition during, 132–34
—Destroy the Four Pests Campaign, 15–16, 191n9, 192n9
—"grain first" policy, 4
—Great Leap Forward, the, 12, 13, 18–19, 50, 133, 191n5, 194n27
—Let a Hundred Flowers Bloom, 12, 200n15
—Reform and Opening Up *(gaige kaifang)*, 2–3, 12, 17, 44, 64
—Three Great Cuttings, 17, 192n11
Politics of Piety, The, 87
Population Bomb, The, 59
population pressure/control, 59–63, 202n31
Portugal, 30
Posey, Darrell, 131
PRC. *See* China, People's Republic of (PRC)

Prestholdt, Jeremy, 155, 156, 215n6
protected species. *See under* wildlife

rain forests. *See* tropical rain forests
Ramirez, Renya, 25
Ramon Magsaysay Award, 147
Red Guards, 11, 12, 33, 43, 92, 195n32
Redstockings (feminist organization), 32*fig6*
Reform and Opening Up *(gaige kaifang)*, 2–3, 12, 17, 44, 64
resistance model (of globalization), 23–25, 78–79, 85–87, 91; indigenous resistance, 23–24, 73, 79, 82, 86, 205n2–3, 206n11
rightful resistance, 143, 145
Roediger, David, 123
Rofel, Lisa, 72
rubber. *See under* plantations

sacred forests. *See* sacred lands
sacred lands, 14, 39, 62, 132–34, 139*fig12*, 139*fig13*, 147–49, 212n13, 213n16–17; Pei Shengji's work on, 127–30
San, 130, 211n9
Santiapillai, Charles, 165–66
Schaller, George, 44, 71, 165, 198n6
scientists. *See* Chinese experts and scientists
Scott, James, 78–79, 85, 86, 205n2, 205n3
shaoshu minzu, defined, 122. *See* ethnic minorities *(shaoshu minzu)*
Shou Bin (Xiao Long villager), 178–79
Sikkink, Kathryn, 22
slash-and-burn agriculture, 200n9; deforestation and, 33–34, 46, 52–53, 58–63, 66–71, 102, 160, 181; during Mao era, 15, 52, 133; valorized, 35, 135–36, 148, 149, 166–67, 201n21; Xu Jianchu's work on, 134–40. *See also* swidden agriculture
Sloping Agricultural Land Technology (SALT), 97
Smil, Vaclav, 5, 190n5
Sobel, David, 57
social forestry, 5, 19, 20*fig2*, 55
socialism
and ethnic formations, 122–23, 133
as globalized formation, 26, 30, 122, 150, 187

Southeast Asian Universities Agroecosystems Network, 128
Southwest Forestry College, 2, 65
Soviet Union, 29–30, 49, 54, 122, 150, 186
Species Survival Network, 165
splittism *(fen lie)*, 126, 203n33
Sri Lanka, 162, 165, 166, 181
subalterns, 23, 82, 85, 86, 113, 114
Subaltern Studies Collective, 85, 206n9
superstition *versus* knowledge, 132–34, 137, 148, 212n14–15
swidden agriculture, 200n9; elephant habitat and, 160, 167, 177, 181–82; during Mao era, 51, 91, 94; negative perceptions of, 67, 81–82, 97, 107; positive perceptions of, 35, 134–37, 141, 148, 149, 166. *See also* slash-and-burn agriculture

Taiwan, 126
Tang Xiyang, 61
tea. *See* plantations
Thailand, 35, 40, 48, 164
The Nature Conservancy (TNC). *See* TNC (The Nature Conservancy)
"The Story of Xishuangbanna" (MacKinnon), 138
Thoreau, Henry David, 10
Three Great Cuttings, 17, 192n11
Tibet, 126
Tibetans, 118, 122, 125–26, 210n33, 211n10
TNC (The Nature Conservancy), 125–26, 128, 143, 205n4
Toledo, Alejandro, 117
transnational work: Chinese experts and scientists, 47–57, 70–72; conservationists, expatriate, 63–72; defined, 42–43, 46–47, 69; electric fences as example of, 161, 167; elephants and, 152–54, 164–66; *versus* global flows, 46, 71, 72; preceding WWF's entry to China, 46–47, 53–63, 70–72
tribal slot, 124, 146, 210n2
Tropical Botanical Garden (Xishuangbanna), 106, 128–29, 132
Tropical Forest Experimental Station, 65

tropical rain forests: deforestation, 33–34, 46, 52–53, 58–63, 66–71, 102, 160, 181; global interest in, 33, 57–58, 59–60, 202n28; population pressure and, 59–63; the West's "discovery" of China's, 42, 44, 129, 198n8
Tsing, Anna, 25
tuzhu ren, 116, 120, 132, 148, 210n1

UNESCO, 128, 133; Man and the Biosphere (MAB) program, 177, 180, 181
United Nations, 119, 150
United States, 5, 14, 22, 26, 29, 47, 53, 55, 97, 123, 137, 142, 186; environmentalism in, 6, 10, 36, 57, 109, 160; indigeneity in, 29, 117, 121, 134, 137
U.S.-led embargo of China, 49, 51, 192n10, 199n5, 201n17
Uyghurs, 118, 210n33, 211n7

Vietnam, 1, 5, 33, 48, 100, 128, 130, 181; China's conflicts with, 49, 91, 101, 216n20; Vietnam War, 29, 31*figs*, 54, 91, 100, 186, 197n42
villagers *(nongmin)*, 213n19; as agents of environmental winds, 5–6, 11–12, 21–22; art of engagement and, 80–85, 88–91, 100–112; Bentley's view of, 97–99; elephants and, 160–62, 168–79; environmental degradation and, 61–63, 67, 69, 126, 136, 138, 202n32; Mao and post-Mao periods, 13, 17–18, 50; peasant resistance, 23–24, 78–79. *See also* indigeneity; indigenous peoples and knowledge

Wages of Whiteness, The, 123
wastelands *(huangdi)*, 3, 15, 49, 50–51, 58, 83
Weapons of the Weak, 79
Welland, Sasha, 25
Westernization, as logic of globalization, 6–7, 24–25, 187
West Wind, the *(Xi Feng)*, 26, 187
Wild Elephant Valley (WWF), 94, 162–64, 169–70, 179, 180*fig14*, 182, 217n28
wildlife
—animal agency, 154–58
—charismatic species, 6, 44, 57, 152, 215n3

—protected/endangered species: harm to or by, 105, 170–71, 180–81, 216n16–19; sensibilities/laws around, 18, 36, 38, 92, 93*fig10*, 164, 168, 173, 177
—trade in animal parts, 48, 152, 183, 191n10
—used in art of engagement, 88–89
—wildlands and (China, 1950s–80s), 15–19
—*See also* elephants; pandas
winds *(feng)*, Chinese metaphor of, 189n3, 191n3, 194n27; describing social transformations, 3, 6–7, 11–15, 21–22, 25–27, 36–37; the global 1960s, as example of, 27–33; Mao's use of, 26, 187. *See also* environmental winds *(huanjing feng)*
women's movement, US, 27–29, 28*fig3*, 31–33, 32*fig6*, 36, 185
World Bank, 10, 46, 66, 83, 118, 119, 125, 211n10
World Bank's Global Environmental Facility, 81
world-making approach (to globalization), 25–26, 194n25
World Trade Organization (WTO), 72
Wu Ling (Xiao Long villager), 78
WWF (World Wildlife Fund): history of, 198n3; Prince Philip's trip to Xishuangbanna, 44–47, 45*fig7*, 95; transnational work by, 46–47, 53–63, 70–72
WWF-China. *See* panda project, Sichuan (WWF); Wild Elephant Valley (WWF); Xiao Long, agroforestry project (WWF)

Xiao Long: history, description, 73–78, 75*fig8*, 79*fig9*, 81, 91–94, 171–73
Xiao Long, agroforestry project (WWF): convergence of project's goals, 68–71; demonstration plot, 101–4; fencing, 161–62; fruit trees, 107–12; goals, 62–63, 66–68, 160, 163; project evaluation, 33–36, 166–67, 196n43; relationship to Nature Reserve Bureau, 95–96; villagers and elephants, 160–62, 168–79; villagers' art of engagement, 80–85, 88–91, 101–12. *See also* Bentley, Jack; MacKinnon, John; Yang Bilun
Xishuangbanna (Banna), 16*fig1*, 34*map2*; ecological, geopolitical, industrial significance of, 48–49, 52–53. *See also* elephants; Nature Reserve (Xishuangbanna); plantations; Tropical Botanical Garden (Xishuangbanna); tropical rain forests; WWF (World Wildlife Fund); Xiao Long
Xu Benhan, 52
Xue Hui, Olivia, 7
Xue Jiru (Hseuh Chi-Ju), 1–3, 55
Xu Jianchu, 134–40, 147, 149, 181

Yanagisako, Sylvia, 72
Yang Bilun, 63–65, 80, 94–97, 101–103, 109, 179, 203n36
Yang Yuanchang, Professor, 41–53, 64–65, 159, 165
Yi, 122
Yin Shaoting, 135, 138, 181, 213n18
Yin Yigong, 52
Yunnan Province, China, 9*map1*, 34*map2*; as hub of global environmentalism, 2, 5–6, 9–10, 14–15, 43–46, 56; impact of environmental policies on, 15–22. *See also* elephants; Nature Reserve (Xishuangbanna); plantations; Tropical Botanical Garden (Xishuangbanna); tropical rain forests; WWF (World Wildlife Fund); Xiao Long
Yu Xiaogang, 140–48

Zapatista movement, 117, 206n11
Zhang Bing (Xiao Long villager), 84
Zhang Li, 170–71
Zhan Mei, 25
Zheng Zuoxin (Cheng Tso-hsin), 42, 43
Zhu Xiang, 165